PRACTICAL SUPPORT
FOR CMMI®-SW
SOFTWARE PROJECT
DOCUMENTATION

IEEE Computer Society Publications
The world-renowned IEEE Computer Society publishes, promotes, and distributes a wide variety of authoritative computer science and engineering texts. These books are available from most retail outlets. Visit the CS Store at *http://computer.org/cspress* for a list of products.

IEEE Computer Society / Wiley Partnership
The IEEE Computer Society and Wiley partnership allows the CS Press authored book program to produce a number of exciting new titles in areas of computer science and engineering with a special focus on software engineering. IEEE Computer Society members continue to receive a 15% discount on these titles when purchased through Wiley or at wiley.com/ieeecs

To submit questions about the program or send proposals please e-mail dplummer@computer.org or write to Books, IEEE Computer Society, 100662 Los Vaqueros Circle, Los Alamitos, CA 90720-1314. Telephone +1-714-821-8380.
Additional information regarding the Computer Society authored book program can also be accessed from our web site at *http://computer.org/cspress*

PRACTICAL SUPPORT FOR CMMI®-SW SOFTWARE PROJECT DOCUMENTATION

Using IEEE Software Engineering Standards

SUSAN K. LAND

JOHN W. WALZ

IEEE
COMPUTER
SOCIETY

A WILEY-INTERSCIENCE PUBLICATION

Published by John Wiley & Sons, Inc., Hoboken, New Jersey
Published simultaneously in Canada.

For general information on our other products and services please contact our Customer Care Department within the U.S. at 877-762-2974, outside the U.S. at 317-572-3993 or fax 317-572-4002.

Wiley also publishes its books in a variety of electronic formats. Some content that appears in print, however, may not be available in electronic format.

Library of Congress Cataloging-in-Publication Data is available.

ISBN-13 978-0-471-73849-7
ISBN-10 0-471-73849-2

Printed in the United States of America.

10 9 8 7 6 5 4 3 2 1

CONTENTS

PREFACE

The IEEE Computer Society Software and Systems Engineering Standards Committee (S2ESC) is the governing body responsible for the development of software and systems engineering standards. S2ESC has conducted several standards users' surveys, the results of which indicated that standards users found the most value in the guides and standards that provided the specific detail they needed for the development of their process documentation. Users consistently said that they used the guides in support of software process definition and improvement (ISO 9001 or CMMI®) but that these standards and guides required considerable adaptation when applied as an integrated set of software process documentation.

This book is written to support software engineering practitioners who are responsible for producing the process documentation, and work products or artifacts, associated with support of software process definition and improvement. It is the hope of the authors that this book will help members of organizations who are responsible for developing or maintaining their software processes in order to support CMMI-SW® (Levels 2 and 3, see reference [61]) documentation requirements.

Software process definition, documentation, and improvement should be an integral part of every software engineering organization. This book addresses the specific documentation requirements in support of the CMMI-SW® by providing detailed documentation guidance in the form of:

- Detailed organizational policy examples
- An *Integrated set* of over 20 deployable document templates
- Examples of over 50 common work products required in support of assessment activities
- Examples of organizational delineation of process documentation

This book provides a set of templates based on IEEE Software Engineering Standards that support the documentation required for all activities associated with software development projects. The goal is to provide practical support for individuals responsible for the development and documentation of software processes and procedures. The objective is to present the reader with an integrated set of documents that support the requirements of the CMMI-SW® Levels 2 and 3. This book is meant to both complement and extend the information provided in the book *Jumpstart CMM®/CMMI® Software Process Improvement Using IEEE Software Engineering Standards* by Susan Land, which provides a detailed mapping of both the CMM® and the CMMI-SW® to the IEEE standards set and provides a logical basis for the material contained within this text.

It is hoped that this book will provide specific support for organizations pursuing software process definition and improvement. For organizations that do not wish to pursue CMMI® accreditation, this document will show how the application of IEEE Standards can facilitate the development of sound software engineering practices.

ACKNOWLEDGMENTS

Susan K. Land would like to acknowledge her employer, Northrop Grumman IT/TASC, and thank them for their continued support of her IEEE Computer Society volunteer activities. She would also like to acknowledge her colleagues within the volunteer organizations of the IEEE Computer Society for their constant encouragement, their dedication to quality, and their consistent pursuit of software engineering excellence. Also, she would like to thank her husband for his unwavering support and encouragement.

John W. Walz would like to acknowledge the support of his wife, without whom this book would not have become a reality—thank you. He would also like to acknowledge his employer, The Sutton Group, and thank Stan Flowers for his continued support of his IEEE Computer Society volunteer activities. Also, he would like to thank his previous managers who guided his professional development: Thomas J. Scurlock, Jr and Terry L. Welsher, both retired from Lucent Technologies; and James R. McDonnell, SBC. For his long involvement in software engineering standards, he would like to acknowledge the leadership of Helen M. Wood.

1

INTRODUCTION AND OVERVIEW

INTRODUCTION

Moving an organization from the chaotic environment of free-form development toward a more controlled and documented process can be overwhelming to those tasked to make it happen. This book is written to support software engineering practitioners who are responsible for producing the process documentation and work products or artifacts associated with support of software process definition and improvement. It specifically addresses how IEEE standards may be used to facilitate the development of processes, internal plans, and procedures in support of managed and defined software and systems engineering processes for CMMI-SW® Levels 2 and 3. It describes how IEEE Software Engineering Standards can be used to help support the development and definition of corporate best practices.

The IEEE Computer Society Software and Systems Engineering Standards Committee (S2ESC) offers an integrated suite of standards based on using IEEE/EIA 12207 as a set of reference processes. S2ESC recommends that organizations should use 12207 and related standards to assist them in defining their organizational processes and then assess those processes using CMMI® or alternative methods. However, this book is based on the premise that the reader's processes may not have been defined in this fashion and that the reader seeks help in improving those processes in anticipation of a CMMI assessment. Therefore, the advice provided in this book does not make the assumption that the user's processes conform to 12207. In general, that advice will be consistent with 12207, but its intent is to assist with process documentation and improvement in the context of the CMMI.

IEEE standards can be used as tools to help with the process definition and documentation (e.g., work products) required in support of the process improvement of software and systems development efforts. Many of the IEEE software engineering standards provide detailed procedural explanations, they offer section-by-section guidance on building the necessary support material and, most importantly, they provide best practice guidance in

Practical Support for CMMI®-SW Software Project Documentation. By S. K. Land and J. W. Walz
© 2006 IEEE Computer Society

support of process definition as described by those from academia and industry who sit on the standards review panels.

The CMMI®-SW [61] is a compendium of software engineering recommended requirements that acts as a motivator for the continuous evolution of improved software and system engineering processes. It is the premise of the information provided within these pages that IEEE Software Engineering Standards can be used to provide the basic beginning framework for this type of process improvement.

The CMMI® does not tell the user how to satisfy process area (PA) criteria. The CMMI® is a descriptive model, describing the criteria that the processes should support. IEEE standards are prescriptive. These standards describe how to full fill the requirements associated with all activities of an effective software and systems project life cycle.

It is often hard to separate the details associated with product development from the practices required to manage the effort. Simply handing the CMMI® to a project lead or manager provides them with a description of an end results model. Pairing this with IEEE standards provides them with a way to work toward this desired end. IEEE standards do not offer a "cookie cutter" approach to management; rather, they support the definition of the management processes in use by describing what is required.

For organizations that do not wish to pursue CMMI® Level 2 accreditation, the CMMI® will show how the application of IEEE standards, and their use as reference material, can facilitate the development of sound software and systems engineering practices. This book is geared for the CMMI® novice, the project manager, and practitioners who want a one-stop source, a helpful document that provides the details and implementation support required when pursuing process definition and improvement. The plans described in this book may be used as an integrated set in support of software and systems process definition and improvement. It should be noted that many of the plans offer examples; these examples are meant to show intent and offer only narrow perspectives on possible plan content.

The authors recommend both attendance in SEI's three-day course, "Introduction to CMMI," and acquisition of the IEEE Software Engineering Collection, which fully integrates over 40 of the most current IEEE software engineering standards onto one CD-ROM. This book does not provide a mapping between the IEEE standards set and the CMMI®. This is provided in the book *Jumpstart CMM®/CMMI® Software Process Improvement: Using IEEE Software Engineering Standards* (Wiley/IEEE Press/IEEE Computer Society; 2005).

WHAT IS THE CMMI®?

The CMMI® is a set of process frameworks. It

- Contains a framework that provides the ability to generate multiple models and associated training and assessment materials. These models may represent:

    Software and systems engineering

    Integrated product and process development

    New disciplines, such as supplier sourcing

    Combinations of disciplines
- Provides guidance to use when developing processes within your tailored model.

WHAT THE CMMI® IS NOT

The CMMI® does not provide processes or process descriptions. Actual processes are dependent upon:

- Application domain(s)
- Organization structure
- Organization size
- Organization culture
- Customer requirements or constraints

WHAT ARE STANDARDS?

Standards are consensus-based documents that codify best practice [42]. Consensus-based standards have seven essential attributes that aid in process engineering. They

- Represent the collected experience of others who have been down the same road
- Describe in detail what it means to perform a certain activity
- Can be attached to or referenced by contracts
- Help to assure that two parties are using the same definition of an engineering activity
- Increase professional discipline
- Protect the business and the buyer
- Improve the product

IEEE software engineering standards provide a framework for defining and documenting software and systems engineering activities. The "soft structure" of the standards set lends itself well to the instantiation of CMMI®-SW Level 2 and 3 process areas. The structure of the IEEE software engineering standards set provides for organizational adaptation. Each standard describes recommended best practices detailing required activities. These standards documents provide a common basis for documenting organizationally unique process activities.

2

SUMMARY OF CMMI-SW®

THE CMM®-SW

The CMM® for Software was finalized in 1991, leading many organizations to ambitiously pursue software process improvement. The higher levels of the CMM® required that organizations institute enterprise-wide change and support. As organizations changed and improved their processes in support of the CMM® for software, other CMM® process models were developed. The CMM® Integration is the consolidation of three source models: The Capability Maturity Model for Software (SW-CMM®) v2.0 draft C [49], the Electronic Industries Alliance Interim Standard (EIA/IS) 731 [81], and the Integrated Product Development Capability Maturity Model (IPD-CMM®) v0.98 [64]. When effectively implemented, the CMMI® supports enterprise cultural change, process definition, and improvement.

CMMI®-SW Continuous Versus Staged

The CMMI® provides for multiple models and two representations: continuous or staged. There are four CMMI® models: Systems Engineering, Software Engineering, Integrated Product and Process Development, and Supplier Sourcing. When implementing CMMI® process improvement, the choice of a selected representation can depend upon a number of factors. The continuous representation allows for the selection of the order of improvement, enables comparisons across organizations by process area, provides for the migration from Electronic Industries Alliance Interim Standard (EIA/IS) 731, and provides for comparison to the International Organization for Standardization and International Electrotechnical Commission (ISO/IEC) 15504 Technical Report.

The staged representation more closely resembles the original CMM®, providing for sequential, progressive process improvement. The staged representation allows for organizational comparison across maturity levels and provides a single rating summarizing appraisal results, thereby providing an easy migration from the CMM® for software.

The continuous representation uses capability levels to measure process improvement. There are six capability levels, numbered 0 through 5. Each capability level corresponds to a generic goal and a set of generic and specific processes, as shown in Table 2-1.

Maturity levels, which belong to the staged representation, apply to an organization's overall maturity. There are five maturity levels, numbered 1 through 5. Each maturity level comprises a predefined set of process areas (see Table 2-2).

Both capability and maturity levels have the same labels; the main difference between maturity levels and capability levels is the process improvement strategy taken, which determines the CMMI representation.

Structural Elements of the CMMI®-SW

The CMMI® has five maturity levels. Level 2 is the level at which the software process is effectively managed. Each of these maturity levels is also supported by specific practices. There are 21 process areas (PAs) distributed over maturity Levels 2–5; there are no PAs at the initial Level 1. Table 2-3 lists the PAs of each maturity level and the number of specific and generic practices. To simply reach Level 2 means that an organization must demonstrate 125 supporting practices!

PROCESS AREAS

In addition to defined levels of maturity, the CMMI® also identifies process areas (PAs). These PAs are grouped into the four categories of Process Management, Project Management, Engineering, and Support.

Table 2-1. CMMI® Continuous representation capability levels [49]

Capability level	Continuous representation capability levels
0	Incomplete
1	Performed
2	Managed
3	Defined
4	Quantitatively Managed
5	Optimizing

Table 2-2. CMMI® Staged representation maturity levels [49]

Maturity level	Staged representation maturity levels
1	Initial
2	Managed
3	Defined
4	Quantitatively Managed
5	Optimizing

Table 2-3. Matrix of CMMI®-SW PAs and supporting practices

Maturity level	Process area (PA)	# of practices
5 Optimizing	Organizational Innovation and Deployment Causal Analysis and Resolution	19 17
4 Quantitatively Managed	Organizational Process Performance Quantitative Project Management	17 20
3 Defined	Requirements Development Technical Solution Product Integration Verification Validation Organizational Process Focus Organizational Process Definition Organizational Training Integrated Project Management Risk Management	20 21 21 20 17 19 17 19 20 19
2 Managed	Requirements Management Project Planning Project Monitoring and Control Process and Product Quality Assurance Configuration Management Supplier Agreement Management Measurement and Analysis	15 24 20 14 17 17 18

Project Management

The Project Management process areas cover the project management activities related to the planning, monitoring, and control of projects. The Project Management Level 2 process areas of CMMI® include Project Planning, Project Monitoring and Control, and Supplier Agreement Management. For Level 3, the two additional Project Management PAs are Integrated Project Management and Risk Management.

Engineering

The Engineering process areas group development and maintenance activities that are shared across engineering disciplines (e.g., systems engineering and software engineering). The CMMI® PA Requirements Management is the Level 2 PA in support of the Engineering process category. For Level 3, the five additional Engineering PAs are Requirements Development, Technical Solution, Product Integration, Verification, and Validation.

Support

The Support process areas concentrate activities supporting product development and maintenance. These Support process areas address performance processes, or those used in support of other PAs. Level 2 processes included in the Support grouping are Configuration Management, Process and Product Quality Assurance, and Measurement and Analysis. For Level 3, the additional Support PA is Decision Analysis and Resolution.

Process Management

The Process Management process areas focus on the organizational activities related to the definition, planning, deployment, implementation, monitoring, control, appraisal, measurement, and improvement of processes. There is no CMMI® Level 2 PA that directly supports this category; rather, this PA category plays a key role in Level 3 process implementation. For Level 3, the three Process Management PAs are Organizational Process Focus, Organizational Process Definition, and Organizational Training.

Specific and Generic Goals

Each maturity level contains process areas, specific goals, specific practices, generic goals, generic practices, typical work products, subpractices, notes, discipline amplifications, generic practice elaborations, and references (see Figure 2.1).

Specific and Generic Practices

In each representation, there are specific and generic goals, which are numbered sequentially. Each specific goal has a number beginning with the prefix SG. Each generic goal has a number beginning with the prefix GG. The practice title is not used for appraisals or rated in any way. However, the practice statement is used for process-improvement and appraisal purposes (see Table 2-4).

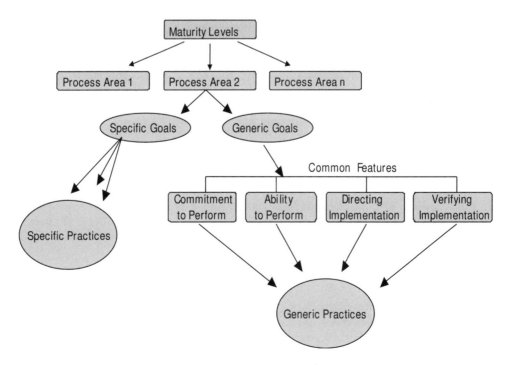

Figure 2-1. Overview of the CMMI®-SW [49].

Table 2-4. Example CMMI® specific practice [61]

SG1	Title: Manage Requirements **Requirements are managed and incorporated with project plans and work products are identified.**
SP 1.1	*Title: Obtain an Understanding of Requirements;* *Develop an understanding with the requirements providers on the meaning of the requirements.*
SP1.2	*Title: Obtain Commitment to Requirements;* *Obtain commitment to the requirements from the project participants.*

Specific practices begin with the prefix SP, followed by a number in the form x.y. The x is the number of the specific goal that the practice is mapped to, and the y is the practice's sequence number. For example, in the Requirements Management process area, the first specific practice associated with specific goal 1 is numbered SP 1.1 and the second is SP 1.2, as shown in Table 2-4.

Generic practices are numbered in a similar way beginning with the prefix GP, followed by a number in the form x.y, where x is the number of the generic goal that the practice is mapped to, and y is the practice's sequence number (see Table 2.5). A second number is used for the generic practices, indicating the sequence number of the practice within one of the four common feature categories to which it belongs. For example, the first generic practice associated with GG 2 is numbered GP 2.1 and CO 1. The CO 1 number indicates that the generic practice is the first generic practice organized under the Commitment to Perform common feature.

Table 2-5. Example CMMI® generic practice [61]

GG2	Title: Institutionalize a Managed Process **The process is institutionalized as a managed process.**
GP 2.1 (CO1)	*Title: Establish an Organizational Policy;* *Establish and maintain an organizational policy for planning and performing the requirements management process.*

CMMI®-SW Common Features

There are four common features used in CMMI® models:

1. Commitment to Perform (CO) identifies the generic practices relating to the creation of policies and corporate sponsorship.
2. Ability to Perform (AB) identifies the generic practices that ensure that a project, or organization, has adequate designated resources.
3. Directing Implementation (DI) defines the group of generic practices relating to the management of process performance, the integrity of associated work products, and stakeholder involvement.
4. Verifying Implementation (VE) designates the group of generic practices relating to management review and conformance evaluation.

These common features provide a basis for organization for the generic practices of each process area. Common features are model components that provide a way to present the generic practices. Each common feature is designated by an abbreviation as shown above.

CMMI®-SW Components

The components of a CMMI® models are grouped into three categories: required, expected, and informative. Table 2-6 provides an example of these three categories.

Required Components

All specific goals and generic goals are required model components. These components must be clearly demonstrated by an organization's processes. Required components are essential to rating the achievement of a process area. Goal achievement (or satisfaction) is used in appraisals as the basis upon which process area satisfaction and organizational maturity are determined.

Expected Components

Specific practices and generic practices are the crucial CMMI® model components. These components clearly define any process-supporting elements and are fundamental to rating the achievement of a process area. Expected components describe what an organization will typically implement to achieve a required component. Expected components guide those implementing improvements or performing appraisals. These components support the required goals for process improvement and are expected to be present in the planned and implemented processes of the organization before goals can be considered satisfied.

Informative Components

Informative components provide details that help model users get started in thinking about how to approach goals and practices. These subpractices, typical work products, discipline amplifications, generic practice elaborations, goal and practice titles, goal and practice notes, and references are informative model components (see Table 2-6.) that help model users understand the goals and practices and how they can be achieved.

Table 2-6. Example of CMMI® component categories

Required	SG 1. Requirements are managed and inconsistencies with project plans and work products are identified.
Expected	SP 1.1. Develop an understanding with the requirements providers on the meaning of the requirements.

Informative	Typical work products
	    Lists of criteria for distinguishing appropriate requirements providers; criteria for evaluation and acceptance of requirements.
	Results of analyses against criteria
	    An agreed-to set of requirements subpractices; criteria established for distinguishing appropriate requirements providers and for the acceptance of requirements.

3

ORGANIZATION INSTITUTIONALIZATION

CHARACTERISTICS OF INSTITUTIONALIZATION

Institutionalization is the routine way of doing business that an organization follows as part of its corporate culture. The institutionalization of a process is critical to the effective implementation of organizational process improvement. A process is considered institutionalized if it is entrenched in the way an organization conducts related activities. The CMMI® generic practices are organized by common features. The generic goals and practices within an organization are those that most strongly contribute to the institutionalization of best practices. Tables 3-1 and 3-2 list generic practices for the institutionalization of both managed and defined processes.

GENERIC PRACTICES

As described by the CMMI®, a generic practice (GP) is a practice that applies to all process areas. However, the application of the generic practice will vary for each process area. For example, the generic practices associated with planning may be implemented differently according to the types of planning required in support of each specific process area.

When examining the CMMI® practices that are common to each of the Level 2 and Level 3 goals, it is important to note that the generic practices follow a numbering scheme that associates the GP with its expected level of implementation. As shown below, GPs that begin with a GP 2 or GP 3 are associated with the staged representation of the CMMI®. The key is to remember that generic practices that begin with a GP 2 must be reflected by organizations claiming CMMI-SW (Staged) Level 2 status. If organizations are implementing the staged model, level 3 criteria may be postponed until the organization attains level 2 status and wishes to pursue maturity levels of 3 or higher.

Table 3-1. Institutionalizing the managed process [61]

2.1	Organizational policies are adhered to
2.2	Established plans and process descriptions are followed
2.3	Adequate resources are provided
2.4	Responsibility and authority for performing the process are assigned
2.5	The people performing and supporting the process are trained
2.6	Work products are placed under appropriate levels of configuration management
2.7	Relevant stakeholders are identified and involved
2.8	The performance of the process is monitored against performance plans and corrective actions are taken when required
2.9	The process, its work products, and its services are evaluated for adherence to the process descriptions, objectives, and standards, and issues of noncompliance are addressed
2.10	All process activities, status, and results are reviewed with higher-level management, and corrective action is taken when required

Table 3-2. Institutionalizing the defined process [61]

3.	All items that institutionalize a managed process are addressed
3.1	The description of the defined process for the project or organizational unit is established
3.2	Work products, measures, and improvement information are derived from information collected from the planning and performance of defined processes

CMMI-SW Level 2 Generic Practices

In order to achieve Generic Goal 2 "Institutionalize a Managed Process," each of the six Process Areas for Maturity Level 2 must demonstrate achievement of ten Generic Practices 2.1 to 2.10.

Organizational Policy

An organization must have a commitment to perform its established processes, and these processes are ingrained and are improved.

GP 2.1 Establish an Organizational Policy; establish and maintain an organizational policy for planning and performing the process.

This generic practice is generally the responsibility of senior management. It should define and describe the organization's overall vision of the process. It should:

- Define organizational expectations
- Provide visibility to those who may be affected by these expectations

Refer to Appendix C, Software Processes Work Products—Organizational Process Definition for several examples of organizational policy statements.

Planning

The purpose of this generic practice in support of process planning is to ensure that an organization has an identified implementation path, adequate resources, and consensus agreement regarding process definition and improvement. This GP reads as follows:

GP 2.2 Plan the Process; establish and maintain the plan for performing the process.

This GP requires that the implementing organization establish a process plan and also demonstrate that this plan is maintained in accordance with any required corrective action. Table 3-3 provides a list of the typical contents of a process plan; Chapter 5, Project Planning, provides additional detailed information in support of this generic practice.

Resources

An organization must demonstrate that adequate resources have been provided in support of software process definition and improvement and that these are the resources that are described by the process plan. It is important to understand that the plan will change and changes in planning should also be reflected in changes in resource allocation. Initial resource estimates may, or may not, be adequate and may change over time. The following is GP 2.3 as described by the CMMI®:

GP 2.3 Provide Resources; provide adequate resources for performing the process, developing the work products, and providing the services of the process.

During planning, resources should not only include funding, but should also include facility requirements, skill levels of individuals, and tools required in support of the process development and management.

Responsibility

This generic practice ensures that individuals have been assigned responsibility for performing the software process. It is important to understand that simple assignment of responsibility is not enough; an organization must also demonstrate that the people assigned are held accountable and that they are also empowered with the authority required for implementation. Generic Practice 2.4 describes the responsibility practice:

Table 3-3. Typical contents of process plan

The process description
Any associated or referenced standards
All support requirements
All performance objectives
A description of any dependencies
Any required resources
The identification of responsibility and authority
Any required training
Related configuration management requirements
Related measurement requirements
The involvement of named stakeholders
All monitoring and control activities
The methodology for objective process evaluation
All management review activities

GP 2.4 Assign Responsibility; assign responsibility and authority for performing the process, developing the work products, and providing the services of the process.

There are many ways that an organization can demonstrate the assignment of responsibility. The incorporation of specific language in process plans, job descriptions, or employee performance evaluations can be used to demonstrate the assignment of responsibility. A more dynamic demonstration of this could be shown in the form of meeting minutes or action items.

Training

Individuals responsible for software process definition, implementation, performance, and improvement must be adequately trained. Organizations must demonstrate that training is provided and that this training is appropriate. This generic practice as described by the CMMI® is:

GP 2.5 Train People; train the people performing or supporting the process as needed.

Organizations have flexibility in the way that training is implemented and this requirement can be fulfilled by both formal and informal training. Training can range from self-study to formal classroom training but must reinforce a common understanding of the software process specific to the organization. Chapter 5, Organizational Training, provides additional detailed information in support of the development of a training plan.

Managing Configurations

When implementing a software development effort, it is essential that all items critical to the success of the effort be under configuration control. It is just as important to place the work products associated with a practice or process under configuration control. The process definition, along with all of its supporting tools and artifacts, should be managed. The GP is as follows:

GP 2.6 Manage Configurations; place designated work products of the process under appropriate levels of configuration management.

Changes in configuration can reflect maturation in the supporting process and the evaluation of process effectiveness. It is important to understand that all items supporting process and product development should be identified and placed under configuration control. Chapter 4, Configuration Management, provides additional detailed information in support of this generic practice.

Stakeholders

Getting the "buy-in" of key participants is critical to the success of any process improvement effort. This GP requires the identification of key stakeholders and their appropriate participation in activities related to the practice. Table 3-4 provides a list of some of the typical target areas for stakeholder involvement. The CMMI® calls out this generic practice as GP 2.7:

Tab le 3-4. Typical target areas
for stakeholder involvement

Planning
Decisions
Communications
Coordination
Reviews
Appraisals
Requirements definitions
Resolution of problems/issues

GP 2.7 Identify and Involve Relevant Stakeholders; identify and involve the relevant stakeholders as planned.

Monitor and Control

Processes must be monitored to ensure that they are being effectively controlled. This practice requires that enough visibility be provided into the process to allow for appropriate corrective action if required. The GP reads:

GP 2.8 Monitor and Control the Process; monitor and control the process against the plan for performing the process and take appropriate corrective action.

Organizations should use the process plan to provide a basis for performance analysis and management. All accomplishments and/or deviations from the process plan should be reviewed with all stakeholders. Corrective action should be taken when it is clear that the plan objectives are not being met or when progress is deviating significantly from that reflected in the planning. Chapter 4, Project Monitoring and Control and Measurement and Analysis, provides additional detailed information in support of this generic practice.

Objective Evaluation of Adherence

In order for process improvement to be effective and for it to provide the intended benefit, it must be objectively evaluated. This practice is designed to determine whether the process is not implemented as planned or does not comply with the described process. The practice states:

GP 2.9 Objectively Evaluate Adherence; objectively evaluate adherence of the process against its process description, standards, and procedures, and address noncompliance.

In order for the evaluation to be truly objective, it is important that individuals not directly responsible for the process evaluate adherence. It is not a requirement for the evaluation to be accomplished by someone outside of the organization. Individuals within the same organization can certainly evaluate adherence, but these individuals should be external to process (project) being evaluated. Chapter 4, Process and Product Quality Assurance, provides additional detailed information in support of this generic practice.

Management Review

This generic practice helps to ensure that all levels of management are informed, that they have the information required to support their decision making, and that they can facilitate corrective action if required. GP 2.10 reads as follows:

GP 2.10 Review Status with Higher-Level Management; review the activities, status, and results of the process with higher-level management and resolve issues.

This type of review can take the form of status reporting or reporting as defined in the supporting plan.

CMMI-SW Level 3+ Generic Practices

In order to achieve Generic Goal 3, "Institutionalize a Defined Process", each process area for Maturity Levels 2 and 3 must demonstrate achievement of Generic Practices 3.1 and 3.2 along with the previous ten for Generic Goal 2. These Generic Practices are incorporated by organizations pursuing the implementation of CMMI®-SW Levels 3, 4, and 5.

Defined Process

For a process to be effective across an organization, a common set of processes must be developed for reference. The IEEE Software Engineering Standards Set [42] provides an ideal reference for practitioners to develop their organizational processes. This is not to say that these standards support a cookie cutter approach to software process definition; rather, these standards provide the norms to which organizations should conform in support of the development of sound software engineering practices and lifecycle methodology.

In order to demonstrate conformance, individual projects should adapt (i.e., tailor—definine how the application of the common process may be changed within limits defined by organizational policy) established organizational processes to the needs of each job. The organizational processes should stand ready to meet all of the requirements of the standards. The organizational processes then may be tailored to the project by omitting processes and activities that are outside the scope of the project. Items may only be omitted when their inclusion does not support the requirements of the effort. For example, suppose your organization has been contracted to develop a piece of software, but not to maintain it. Therefore, you omit the organizational maintenance process and keep the organizational development process. The common set of organizational processes has to be adapted to the particular needs of each project. An organization needs to describe the permissible extent of adaptation and how the adaptation is to be performed. So, in addition to the definition of processes, each organization must also provide tailoring guidelines.

IEEE/EIA 12207, IEEE Standard for Information Technology—Life Cycle Processes, provides a common framework for the development and/or management of software. Most of the other IEEE software engineering standards provide users with insight into recommended software engineering best practices. This combination of a common framework along with the supporting process guidance and can act as a catalyst when used effectively in support of software process definition.

GP 3.1 Establish a Defined Process; establish and maintain the description of a defined process.

This generic practice is closely related to the Organizational Process Definition process area. Chapter 5, Organizational Process Definition, provides additional detailed information in support of this generic practice.

Collect Information

Organizations must have a clear picture of the effort and activity associated with a project. In order to understand how some process may be improved, you must first understand what is actually happening. This generic practice requires the collection of the information associated with the planning or performance of the associated process. All artifacts, including planning information, performance measurements, and all lessons learned must be kept in the organization's process asset library and placed under configuration management controls.

GP 3.2 Collect Improvement Information; collect work products, measures, measurement results, and improvement information derived from planning and performing the process to support the future use and improvement of the organization's processes and process assets.

This generic practice is closely related to the Organizational Process Definition process area. Chapter 5, Organizational Process Definition, provides additional detailed information in support of this generic practice.

4

IMPLEMENTATION GUIDANCE

IEEE SUPPORTED PROCESS IMPROVEMENT

The IDEAL (initiating, diagnosing, establishing, acting, and learning) model is an organizational improvement model that was originally developed to support the CMM®-based software process improvement model. It serves as a roadmap for initiating, planning, and implementing improvement actions. This model can serve to lay the groundwork for a successful improvement effort (initiate), determine where an organization is in reference to where it may want to be (diagnose), plan the specifics of how to reach goals (establish), define a work plan (act), and apply the lessons learned from past experience to improve future efforts (learn). This model (refer to Figure 4-1) serves as the basis for both the SW-CMM® and CMMI® process improvement methodologies. IEEE software engineering standards can provide valuable support during the establishing and acting portions of this model. The information contained within the software engineering standards set can support the creation, testing, refinement, and implementation of development processes.

Define and Train the Process Team (Initiate)

A process improvement team is a chosen group of people who are given responsibility and authority for improving a selected process in an organization; this team must have the backing of senior management. Process owners are responsible for the process design, not for the performance, of their associated process areas. The process owner is further responsible for the process measurement and feedback systems, the process documentation, and the training of the process performers in its structure and conduct. In essence, the process owner is the person ultimately responsible for improving a process. IEEE software engineering standards provide valuable support to the process team and each individual process owner. The standards can be used to help define and document the initial baseline of recommended processes and practices.

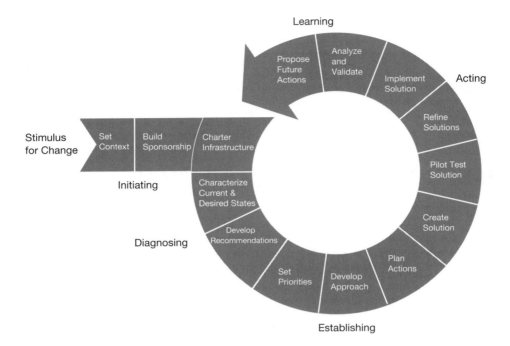

Figure 4-1. The IDEAL model [61].

Implementing process improvement can be very time-consuming, depending upon the scope and complexity of the process. Expectations for the process owner's time commitments and job responsibilities must be modified accordingly to reflect the new responsibilities. This commitment should reflect time budgeted for process definition and improvement and any required refresher training.

Software Engineering

Practically defined, software engineering is a discipline whose goal is the production of fault-free software, delivered on time and within budget, meeting customer expectations. IEEE Std 610.12, the Standard Glossary of Software Engineering Terminology, defines Software Engineering as:

> (1) The application of a systematic, disciplined, quantifiable approach to the development, operation, and maintenance of software; that is, the application of engineering to software.
> (2) The study of approaches as in (1) [2].

Traditionally, individuals graduating with computer science degrees have not been trained in a manner meeting this definition. In fact, except for universities that offer senior-level undergraduate software engineering courses or master's degrees in software engineering, computer science students have been typically trained in the various programming methodologies. Emphasis is too frequently placed upon the "craft" of programming rather than the processes in support of software engineering. IEEE software engineering standards can be used as supplemental training material for practitioners. These standards distill and present the consensus of software engineering best practices.

Graduates are called "computer scientists," but they are often performing engineering job functions without the benefit of formal engineering training. Many might argue that software engineering is a clearly defined profession and recent graduates possess all the required skills in support of both programming and process. If this is a true statement, then why does the software process vary so drastically from organization to organization?

In order for software engineering to be truly defined as a profession, several questions need to be answered: What is software engineering's core body of knowledge? How should software engineers become trained? And how should software engineers become certified? These are the types of questions that can be answered by those practicing in other professional fields.

SWEBOK

On May 21, 1993, the IEEE Computer Society Board of Governors approved a motion to "establish a steering committee for evaluating, planning, and coordinating actions related to establishing software engineering as a profession" [44]. Many professions are based on a body of knowledge that supports such purposes as accreditation of academic programs, development of education and training programs, certification of specialists, or professional licensing. One of the goals of this motion was to establish a body of knowledge in support of software engineering. In 1996, the initial straw man version of this body of knowledge was published as the Software Engineering Body of Knowledge (SWEBOK®*). The 2004 version of the SWEBOK® has been released and is available at www.swebok.org.

The SWEBOK® outlines the knowledge associated with what has become the consensus body of software engineering knowledge. The goal of the SWEBOK® is to define the core of what the software engineering discipline should contain. The acknowledgement of the SWEBOK® by educational institutions would mean a shift from technology-specific programming courses, like those concentrating on C++ and Java, to the knowledge and practices required in support of software development and software project management. The SWEBOK® is divided into ten areas (refer to Table 4-1) that contain descriptions and references to topically supportive material that reflect the knowledge required in support of the software engineering discipline. Process teams can use the SWEBOK® as a reference tool supporting their requirements for process development and definition.

The SWEBOK® will continue to evolve, as do all the bodies of knowledge associated with other professional disciplines. Areas will be redefined and refined over time to address all of the factors that affect the creation or maintenance of a software product. The SWEBOK® offers a challenge to all of those involved with software—to broaden the knowledge base and move from computer science to software engineering. The requirement is for software engineering practitioners to not only know how to write code, but to understand all aspects of the processes that support the creation of their products.

IEEE and Software Engineering Training

In order for software process improvement to be successful, everyone involved should know how to effectively perform his or her roles. Many times, key individuals have not

---

*SWEBOK is a registered trademark of the IEEE.

Table 4-1. Ten SWEBOK areas [44]

SWEBOK® area	Supports
Software requirements	Software requirements fundamentals Requirements process Requirements elicitation Requirements analysis Requirements specification Requirements validation Practical considerations
Software design	Software design fundamentals Key issues in software design Software structure and architecture Software design quality analysis and evaluation Software design notations Software design strategies and methods
Software construction	Software construction fundamentals Managing construction Practical considerations
Software testing	Software testing fundamentals Test levels Test techniques Test-related measures Test process
Software maintenance	Software maintenance fundamentals Key issues in software maintenance Maintenance process Techniques for maintenance
Software configuration management	Management of the SCM process Software configuration identification Software configuration control Software configuration status accounting Software configuration auditing Software release management and delivery
Software engineering management	Initiation and scope definition Software project planning Software project enactment Review and evaluation Closure Software engineering measurement
Software engineering process	Process implementation and change Process definition Process assessment Product and process measurement

Software tools and methods	Software tools
	   Software requirements tools
	   Software design tools
	   Software construction tools
	   Software testing tools

Table 4-1. Ten SWEBOK areas [44]

SWEBOK® area	Supports

Software tools and methods (*cont.*)	Software tools (*cont.*)
	  Software maintenance tools
	  Software engineering process tools
	  Software quality tools
	  Software configuration management tools
	  Software engineering management tools
	  Infrastructure support tools
	  Miscellaneous tool issues
	Software engineering methods

Software quality	Software quality fundamentals Software quality management processes Practical considerations

been effectively trained to support the software engineering process. Table 4-2 provides a list of sample questions that can help determine the level of software engineering expertise an individual may possess.

If people in your organization cannot answer these questions right away or if nobody knows what these roles are or who performs them, then your staff has an urgent need for some software engineering education. Individuals on the process team must be equipped with the appropriate knowledge prior to beginning software process definition and improvement activities.

IEEE software and system engineering standards provide process and practice support to the software engineering knowledge as defined by the SWEBOK®. Each standard contains additional reference material that may be used in support of software engineering training activities. Many of the specific questions that practitioners have may be answered by turning to the IEEE Software Engineering Standards Collection. Table 4-3 provides a list of knowledge requirements and the IEEE software engineering standard that may be used to support the requirement.

Table 4-2. Key software engineering questions

To the project managers:
  a. What is the difference between a plan and a schedule?
  b. What do you record about the estimates that are being made?
  c. Do you estimate size as well as effort when doing your planning? Do you monitor both attributes during the life of the project?

To the configuration managers:
  a. What is a baseline?
  b. What is the purpose of a configuration audit?
  c. Who authorizes changes to the configuration units?

To the quality assurance analysts:
  a. What is the object of quality assurance?
  b. How is it different from quality control? From testing?
  c. Who in the organization knows about the quality assurance activities and results?

Table 4-3. IEEE standards and training

Requirements	
How to document requirements	IEEE Std 830-1998, Recommended Practice for Software Requirements Specification
Test	
How to classify software anomalies	IEEE Std 1044-1993 (R2002), Standard Classification for Software Anomalies
How to select and apply software measures	IEEE Std 982.1-1988, Standard Dictionary of Measures to Produce Reliable Software
How to define a test unit	IEEE Std 1008-1987 (R2002), Standard for Software Unit Testing
What documentation is required in support of the testing process?	IEEE Std 829-1998, Standard for Software Test Documentation
Maintenance	
What maintenance activities are required prior to product delivery?	IEEE Std 1219, Standard for Software Maintenance
Communication	
How to communicate using consistent terminology	IEEE Std 610.12-1990 (R2002), Standard Glossary of Software Engineering Terminology
Configuration management	
What describes the requirements and categories of information in support of configuration management planning?	IEEE Std 828-1998, Standard for Software Configuration Management Plans
Reviews and audits	
Where to find information describing software audit procedures	IEEE Std 1028-1997 (2002), Standard for Software Reviews

Set Realistic Goals (Diagnose)

The CMMI® provides the building blocks for higher maturity at each stage. It is important to set realistic goals when beginning down the path of process improvement. The leap from chaos (Level 1) to Level 2 is often the hardest step for many organizations. Defining the initial process baseline is key to understanding where the organization needs to be; it must first understand where it is. Table 4-4 provides a list of the processes essential to each of the four CMMI areas of focus and may be used to help determine if essential processes are missing or are incomplete.

Use the CMMI®-SW Level 2 and Level 3 goals to identify areas of weakness or bottlenecks in existing processes. Then refer to each of the appropriate IEEE software engineering standards, using them as planning tools and as checklists to be considered when determining how to accomplish process completeness.

It is important to identify which organizational process plans will be developed and the sequence of their development. Many organizations begin with the definition of their SCM policies and practices. Beginning with SCM allows organizations to realize tangible, immediate results. This can provide momentum for the further development of the processes supporting software project planning, software requirements management, software measures definition, SQA, and processes supporting software verification and validation.

Table 4-4. Determine if essential processes are missing or are incomplete [73]

Process management	Engineering
Organizational process focus	Requirements management
Organizational process definition	Requirements development
Organizational training	Technical solution
Organizational process performance	Product integration
Organizational innovation and deployment	Verification
	Validation

Project management	Support
Project planning	Configuration management
Project monitoring and control	Process and product quality assurance
Supplier agreement management	Measurement and analysis
Integrated product management	Decision analysis and resolution
Risk management	Organizational environment for integration
Integrated teaming	Causal analysis and resolution
Integrated supplier management	
Quantitative project management	

Fix Timelines (Establish)

Define your adoption strategy. This is the strategy used by the organization to facilitate process institutionalization. This strategy should include targeted projects, key personnel, training, and, most importantly, schedules reflecting specific software process improvement targets. Goal-driven process improvement is the most effective. Identify both short- and long-term goals stating concise objectives and time periods and use these goals as schedule milestones. Table 4-5 provides a sample timeline for goal-driven process improvement.

Baseline and Implement Processes (Act)

Defining a process baseline is critical when implementing software processes that can be repeatable. Use IEEE standards to develop your baseline process documentation that addresses the requirements of the Level 2 and Level 3 goals. It may be helpful to define the initial process baseline at different levels of abstraction. One useful way of classifying these levels of abstraction is provided by the Basili [70] levels of abstraction:

Reference level with processes and responsibilities defined

Contextual level describing connections among the processes to transfer information

Implementation level combining the mapping of processes to parts of the organization

Detailed information in support of this type of abstraction methodology and the supporting IEEE software engineering standards may be found in *Road Map to Software Engineering—A Standards-Based Guide* [97]. This additional information in support of process abstraction can be useful when trying to define a process baseline.

It is also important to evaluate and identify any potential tools that may be used in support of process automation, keeping in mind that while a tool is not a substitute for a process, it can facilitate its implementation. An ideal candidate area for this type of au-

Table 4-5. Example Goal-driven implementation timeline

(0–3 months)
- Identify individuals responsible for software process improvement.
- Identify project managers who will be participating.
- Identify list of candidate projects.
- Solidify backing of senior management.
- Look at existing processes and make sure they are appropriate and reflect current business needs (small vs. large projects) using CMMI® goals and IEEE software engineering standards.
- Define the formats for your process plans (software configuration management plan, software requirements management plan, software quality assurance plan) using IEEE software engineering Standards and measure them against the CMMI® requirements.
- Get project members to provide feedback on process plans, review and incorporate feedback.
- Conduct ARC Class C Gap Analysis.

(3–6 months)
- Create process document templates for project documentation based upon defined processes; projects will use these to develop their own plans (e.g., software development plan, software requirements specification.)
- Conduct weekly/monthly status reports/reviews to gauge and report progress and provide areas for improvement.

(6–9 months)
- Conduct CMMI®-based reviews of the projects. It would be ideal to also include members from unselected projects to participate in these reviews, along with reporting senior management.
- Provide feedback regarding project review providing requirements for improvement to the projects.

(9–12 months)
- Conduct internal assessments; report to senior management.
- Provide feedback regarding project review, providing requirements for improvement to the projects.

---

tomation is SCM. There are a number of widely marketed tools that support the documentation and control of various types of software configuration items.

Once a process baseline has been established, an action plan should be formulated. The action plan provides a map of the path forward toward improvement of the baseline processes. Table 4-6 provides a sample from an action plan.

Take advantage of the information provided by the IEEE Software Engineering Standards Collection. Many of these standards provide documentation templates and describe in detail what individual project support processes should contain. Think of the standards as an in-house software process consultant who has recommended, based upon years of experience, the proper methodologies and techniques to be used in support of software development.

Perform Gap Analysis (Learn)

It is important to gauge how effectively process improvements have been implemented for continuous process improvement to be successful. This can be determined through the development of a benchmarking appraisal to support gap analysis activities. This type of appraisal will provide a baseline for future process improvement efforts and will identify weaknesses and strengths.

Table 4-6. Example action plan

Process area	Weakness or area for improvement	Short description of how to address	Project point of contact	Resolution date
Project montoring and control	SG2. Although corrective actions are generally tracked to closure, the effectiveness of corrective actions is not consistently addressed.	Create an analysis procedure. This procedure must include a review of the results and effectiveness of corrective actions.	Jim Smith	

To begin, review the associated appraisal methodology that is used in support of the CMMI®. Use these requirements, in conjunction with the CMMI-Staged (SW) Level 2 and Level 3 criteria to develop assessment matrixes. These matrixes can be used to determine model compliance and identify specific deficiencies in implemented practices, or the support documentation that was developed using the IEEE Software Engineering Standards Collection. Table 4-7 provides a list of the CMMI appraisals that may be used in the development of assessment matrixes.

Table 4-7. CMMI® appraisals

Process framework	CMMI®
Appraisal framework	ARC
Assessment	SCAMPI
Capability evaluation	SCAMPI

Perform Self-audit Using CMMI®-SW (Staged) PAs

The components of the CMMI® can be used to form the basis for a gap analysis. These key components are described in detail in the SEI technical report, *Capability Maturity Model Integration (CMMI), Version 1.1* [61]. In this document, each required, expected, and informative component is addressed in detail. Supporting information is also supplied for each specific and generic practice. Use this detailed information when developing assessment matrixes. These matrixes can be used to identify areas of compliance, noncompliance, or areas needing improvement. Table 4-8 provides sample data from a compliance matrix.

IMPLEMENTATION PITFALLS

The implementation of the CMMI®-SW is frought with some common pitfalls. It is important to remember that this model is not prescriptive. Too often, organizations implement process control and improvement, spending considerable time and effort without realizing significant improvements in product cost, quality, or cycle time. In order for

Table 4-8. Example of a CMMI®-SW (Staged) compliance matrix

Req #	Source	Requirement	Satisfied by	Status (P/F/NI)
1	RM SP1.2	How are commitments from participants obtained?	Impact assessments; documented commitment and requirements changes	Pass, fail, or needs improvement (with comments)
2	RM SP1.3	How are requirements managed as they evolve during the project?	Requirements status reporting; requirements tracking database	

organizations to fully benefit from process improvement activities, time must be spent focusing on the definition of current practices and procedures. Only after this initial process definition is complete should organizations move forward to begin to improve their existing processes.

Being Overly Prescriptive

Implementing the recommended practices verbatim can be costly and may not reflect the specific process requirements of an organization. This approach can result in an overly prescriptive process that will increase cost and slow product cycles. This can rapidly destroy the credibility of the process improvement implementation with management and the software development staff. Carefully analyze what areas of process improvement will provide the most positive impact on existing programs, implement these first and management will see the tangible results and continue their support. Small projects may require less formality in planning than large projects, but all components of each standard should be addressed by every software project. Components may be included in the project-level documentation, or they may be merged into a system-level or business-level plan, depending upon the complexity of the project.

Remaining Confined to a Specific Stage

Be careful to not examine the implementation of each stage in isolation. There is certainly room to introduce a few higher maturity level practices early on if there is a good business case for them. For example, verification and validation are Level 3 PAs but an organization can gain significant product improvement and insight into their development process with their implementation during Level 2. Also, many organizations that are just beginning a process improvement effort often delay implementation of a risk assessment program. Risk avoidance can be a critical factor during software development and an organization can significantly benefit if this PA is addressed early on.

Also, be aware that projects within the same organization may be at different maturity levels. In the strictest application of the staged representation, the organization would require that projects practice at maturity Level 2 before moving any of the projects to Level 3. This can take a very long time and is counterproductive. Most organizations identify several pilot projects and move them through the maturity levels. The experience gained from this type of implementation is then carried throughout the entire organization and applied to the remaining efforts.

The CMMI® offers a continuous representation as an alternative to the traditional staged representation, but many organizations are intimidated by the complexity of the models and they are reluctant to make the transition.

Documentation, Documentation

Be careful not to generate policies and procedures to simply satisfy the CMMI® model requirements. Generating documentation for the sake of documentation is a waste of time and resources. Policies and procedures should be developed at the organizational level and used by all projects, only requiring documentation where there is deviation from the standard. The pairing of IEEE Standards, which are prescriptive and describe software engineering minimums, with these models can reduce this risk significantly.

Lack of Incentives

The CMMI® talks about an Engineering Process Group (EPG), reflecting its broader focus. Many times, the success of the process improvement initiatives rests on the shoulders of this group. The EPG should act as a mentor, guiding the software projects through continuous process improvement. Each selected project should be required to follow the recommendations of this group. Process improvement is more effective and the results more permanent when the projects are stakeholders in the software improvement process.

No Metrics

If your organization has effectively implemented process improvement, you can realize improved business results. However, too often organizations implement process improvement just to meet the process goals for certification purposes. Inefficient implementation will e result when organizational needs are not made a priority. Process performance can be static or even degrade relative to business goals. The models do not require substantial process performance measurements until the higher maturity levels. Without a solid understanding of the cost/benefit ratio and its relationship to business results, it can be easy to lose corporate support.

CONCLUSION

Process improvement can be intimidating. Many times, the task of process improvement comes in the form of a directive from senior management, or as a customer requirement, leaving those assigned with a feeling of helplessness. However, all those practicing as software engineers should desire to improve themselves and leave behind the chaotic activities associated with the uncontrolled software processes and required heroic efforts of a Level 1 organization. At the managed level (Level 2), software engineering processes are under basic management control and there is an established management discipline that provides benefits to all involved. When CMMI is used in conjunction with IEEE software engineering standards, the customer may be assured of a lower risk of failure, the organization is provided with accurate insight into the effort, management can more effec-

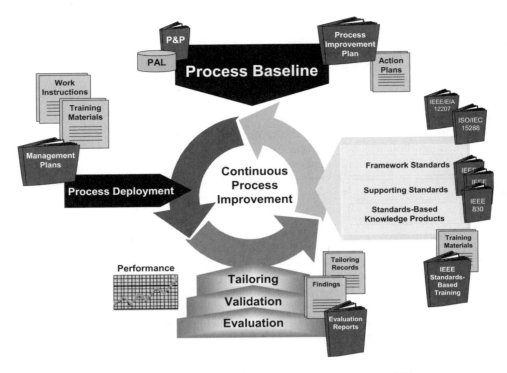

Figure 4-2. Standards Support of Continuous Process Improvement [74].

tively identify and elevate development issues, and team members can work to efficiently managed baselines.

The Level 2 and 3 requirements of the CMMI®-SW are broad. No single IEEE standard can be used in isolation to support these requirements. Rather, a subset of the available IEEE software engineering standards should be employed in combination (refer to Figure 4-2) to provide effective support for Level 2 and 3 CMMI® activities. IEEE software engineering standards can be used to provide detailed, prescriptive support for CMMI® process definition and improvement activities.

5

CMMI®-SW LEVEL 2 SUPPORT

REQUIREMENTS MANAGEMENT

The CMMI®-SW emphasizes bidirectional traceability. The CMMI® requires that a plan for performing the requirements management process be established and that all stakeholders be identified according to this plan. IEEE Standard 830, IEEE Recommended Practice for Software Requirements Specification [6], provides detailed guidance in support of the development of a software requirements management plan. This standard provides detail on what is required to effectively manage software requirements and how to document these requirements in a management plan.

IEEE Std 12207.0, Standard for Information Technology—Software life cycle processes [39], describes seventeen processes spanning the entire life cycle of a software product or service. Even if an organization's processes were defined using other sources, the standard is useful in characterizing the essential characteristics of these software processes and should be considered prior to the implementation of process improvement activities. Referencing this standard, and reviewing what is required for each of these primary process areas, can provide valuable additional guidance in support of the activities associated with requirements management and analysis. It is important that IEEE 12207 be considered prior to the implementation of process improvement activities associated with requirements management and requirements analysis.

CMMI®-SW Goals

The purpose of requirements management is to manage requirements associated with a project and to identify inconsistencies between the requirements and the project plan and associated work products. The CMMI® defines the requirements for the bidirectional traceability of software requirements and specifies that for requirements management to be effective it must operate in parallel with requirements development that provides adequate support for requirements change. The CMMI®-SW also specifically requires that the requirements management process be institutionalized as a managed process.

SG1. Manage Requirements
Requirements are managed and incorporated with project plans and work products are identified.

Specific support for this goal is provided by IEEE Standards 830 [6], 1233 (developing system requirements specifications) [25], IEEE Std 12207, Standard for Information Technology. Software Life Cycle Processes [39], and 1219 (Software Maintenance) [22]. The IEEE Software Engineering Standards Collection supports the idea of requirements-driven software development and/or maintenance activities.

GG2. Institutionalize a Managed Process
The process is institutionalized as a managed process.

The implementation of IEEE Std 830, Recommended Practice for Software Requirements Specifications, assures the development of a software requirements specification but does not support the requirement of the institutionalization of software requirements management. At a minimum, IEEE Std 830 must be used in conjunction with IEEE Std 1058, Standard for Software Project Management Plans [15], and IEEE Std 12207.0, Standard for Information Technology—Software Life Cycle Processes in support of this goal. Incorporating requirements management into the project planning, organizational processes, and management helps to ensure that it is integrated into the lifecycle of the project. Table 5-1 describes the specific and generic goals and practices, typical work products, and commonly associated plans or artifacts in support of the PA.

Software Requirements Management Plan

In order for the IEEE Std 830, IEEE Recommended Practice for Software Requirements Specifications [6], to be an effective instrument for the implementation of Level 2 PAs, information regarding the management of the defined requirements needs to be added and stated explicitly. Information needs to be included in a software requirements management plan (SRMP) that will demonstrate that the resources, people, tools, funds, and time have been considered for the management of the requirements that are defined.

   This information should relate directly to the SPMP. For example, if one person is required to perform the duties of the requirements manager, this should be identified in the SPMP. This section could reference the SPMP if the information is contained in that document. However, the present IEEE standard for SPMPs does not require the identification of requirements management issues such as resources and funding, reporting procedures, and training.

   The recommended SRMP table of contents to support the goals of the CMMI® more directly is shown in Table 5-2.

Software Requirements Management Plan Document Guidance

The following provides section-by-section guidance in support of the creation of a software requirements management plan. The development of an organizational SRMP not

Table 5-1. Requirements management goals and practices [61]

Specific and generic goals/practices and typical work products	Plan or artifact
SG1. Manage Requirements	
SP1.1. Obtain an understanding of requirements	
List of criteria for distinguishing appropriate requirements providers	Requirements management plan
Criteria for evaluation and acceptance of requirements	Requirements management plan
Results of analyses against criteria	Requirements specification review
An agreed-to set of requirements	Requirements specification
SP1.2. Obtain commitments to requirements	
Requirements impact assessments	Requirements specification
Documented commitments to requirements and requirements changes	Signed requirements specification or approved change requests
SP1.3. Manage requirements changes	
Requirements status	Requirements database, baseline change request
Requirements database	Requirements database
Requirements decision database	Requirements database
SP1.4. Maintain bi-directional traceability of requirements	
Requirements traceability matrix	Requirements specification and database
Requirements tracking system	Requirements database
SP1.5. Identify inconsistencies between project work and requirements	
Documentation of inconsistencies including sources, conditions, and rationale	Requirements specification review
Corrective actions	Revised requirements specification
GG2. Institutionalize a Managed Process	
GP2.1. Establish an organizational policy	
Organizational policy supporting requirements management	Organizational policy
GP2.2. Plan the process	
Documentation in support of the requirements management planning process	Requirements management plan
GP2.3. Provide resources	
Identification of resources used in support of requirements identification and management	Software project management plan, requirements management plan
GP2.4. Assign responsibility	
Description of individuals responsible for requirements management activities	Requirements management plan, software project management plan, or organizational policy
GP2.5. Train people	
Identify how individuals will be provided the appropriate training	Training plan or software project management plan

(*continued*)

Table 5-1. *Continued*

Specific and generic goals/practices and typical work products	Plan or artifact
GG2. Institutionalize a Managed Process (*cont.*)	
GP 2.6. Manage configurations	
Description of how requirements and all associated artifacts are placed under configuration management	Configuration management plan
GP 2.7. Identify relevant stakeholders	
Identify who is responsible for the definition and management of requirements	CCB meeting minutes, traceability reporting, and requirements reviews
GP 2.8. Monitor and control the process	
How requirements volatility will be reported	Software measurement and metrics plan, requirements management plan, or software project management plan
GP 2.9. Objectively evaluate adherence	
Describe how adherence is evaluated	Software quality assurance plan
GP 2.10. Review status with higher-level management	
Description of reviews and reporting	Software project management plan
GG3. Institutionalize a Defined Process	
GP 3.1. Establish a defined process	
Establish and maintain a description of the requirements	Requirements management plan management process
GP 3.2. Collect improvement information	Requirements management plan, software measurement and metrics plan, or software project management plan

only helps baseline the process of software requirements definition and management, but the SRMP can also provide valuable training and guidance to project performers. This guidance should be used to help define a management process and should reflect the actual processes and procedures of the implementing organization. Additional information is provided in the document template, *Software Requirements Management Plan.doc,* which is located on the CD-ROM accompanying this book. Additional information in support of requirements traceability is provided in Appendix C, Software Process Work Products.

Purpose. This section should describe the purpose of the software requirements management plan and why effective requirements management is critical to the success of any software development or maintenance effort.

Software requirements management activities are concerned with two major areas:

1. Ensuring that all software requirements are clearly defined
2. Preventing the requirements process from becoming burdened by design, verification, or management details that would detract from specifying requirements

Table 5-2. Software requirements management plan document outline

Title Page
Revision Page
Table of Contents
 1.0 Introduction
 2.0 Purpose
     2.1 Scope
     2.2 Definitions
     2.3 Goals
 3.0 References
     3.1 Key Acronyms
     3.2 Key Terms
     3.3 Key References
 4.0 Management
     4.1 Organization
     4.2 Tasks
     4.3 Responsibilities
         4.3.1 Management
         4.3.2 Program Manager
         4.3.3 Project Lead
         4.3.4 Team Members
         4.3.5 Customer
 5.0 Software Requirements Management Overview
     5.1 Software Requirements Modeling Techniques
         5.1.1 Functional Analysis
         5.1.2 Object-Oriented Analysis
         5.1.3 Dynamic Analysis
     5.2 Software Requirements Management Process
         5.2.1 Requirements Elicitation
         5.2.2 Requirements Analysis
         5.2.3 Requirements Specification
     5.3 Characteristics of a Good SRS
         5.3.1 Correct
         5.3.2 Nonambiguous
         5.3.3 Complete
         5.3.4 Verifiable
         5.3.5 Consistent
         5.3.6 Modifiable
         5.3.7 Traceable
         5.3.8 Usable During Operation and Maintenance
 6.0 Standards and Practices
 7.0 Software Measurement and Metrics
 8.0 Verification and Validation
 9.0 Software Configuration Management
10.0 Developing a Software Requirements Specification
Appendix A // Project Software Requirements Specification Template
Appendix B// Requirements Traceability Matrix

Scope. This section should address the scope of the plan and its applicability.

Definitions. This section should list all relevant defined terminology.

Goals. This section should identify all organizational goals associated with software requirements management.

References. A list of key list acronyms, definitions, and references should be provided in this section of the document.

Organization. This section should describe the individuals responsible for software requirements management. The following is provided as an example:

> The [PROJECT ABBREVIATION] program manager will ensure completion of Software Requirements Management by coordinating the [PROJECT ABBREVIATION]'s team development of an SRS. The program Manager will ensure that the training and time is provided to the project leader to have the SRS properly developed. The SRS will be developed to ensure that requirements are properly captured.

Tasks. This section should describe the high-level tasks associated with software requirements management and the specification of requirements, the following is provided as an example:

> The tasks to be performed in the development of an SRS are:
>
> 1. Specifying software requirements as completely and thoroughly as possible
> 2. Defining software product functions
> 3. Defining design constraints
> 4. Creating software Requirements Traceability Matrix
> 5. Defining assumptions and dependencies

Responsibilities. This section should describe the responsibilities of those individuals at all levels within the organization. It should address the roles of senior management, program and project leaders, team members, and the customer.

Software Requirements Management Overview. This section should provide an overview of the software requirements management process. The following is provided as an example:

> Software requirements management is the set of processes (activities and tasks) and techniques used to define, analyze, and capture the customer's needs for a software-based system. This process starts with the customer's recognition of a problem and continues through the development and evolution of a software product that satisfies the customer's requirements. At that point, the software solution is terminated. For requirements management to work effectively, the requirements must be documented and agreed to by both the customer and [COMPANY NAME]. The formal document that contains this agreement is an SRS. The focus of the requirements management effort is then to create, approve, and maintain an SRS that meets and continues to meet the customer's needs.

Software Requirements Modeling Techniques. This section should provide a description of the software requirements modeling techniques.

Software Requirements Management Process. This section should provide a detailed description of the software requirements management process, the following information is provided as an example:

> The process of creating and approving an SRS can consist of three activities: requirements elicitation, requirements analysis, and requirements specification. Some problems are so clearly stated that only requirements specification is needed, other problems require more refinement and therefore also require the elicitation and analysis. This description will cover all three activities, assuming a worst-case scenario.
>
> Projects following this plan will assess the state of their requirements and enter the process at the appropriate point. (Include a figure that represents the requirements management process with all key players identified at appropriate entry points.)

Requirements Elicitation. Requirements elicitation is the process of extracting information about the product from the customer. Requirements elicitation is highly user interactive. A series of activities may be used to help both the customer and the developer create a clear picture of what the customer requires. These activities are discussed below. They include interviews, scenarios, ambiguity resolution, context analysis, rapid prototyping and clarification and prioritization. This particular ordering of techniques is most useful for moving from the most obscure request to the most definitive. Other orderings also will work well and should be adapted to the project fit.

Interviews. Interviews are the most natural form of gaining an understanding of the customer's request. However, interviews are often performed with little advanced preparation. This is counterproductive. The customer becomes frustrated at being asked unimportant questions; the developer is forced to make assumptions about what is not understood. One way to reduce the failure rate of interviews is to focus the interview using the following strategy:

1. Lose your preconceptions regarding the request. Assuming that you know what must be accomplished prevents you from listening carefully to the customer and may give the customer the perception that the interview is just a formality.

2. Prepare for the interview. Find out what you can about the customer, the people to be interviewed, and the request in general. This gives you a background for understanding the problem. Clarify those areas of uncertainty with the customer, this may help to better formulate requirements.

3. Ask questions that will help you understand the request from a design perspective. These questions lead to clarification of for whom the project is being built and why it is being built. This is similar to the TV commercial that says "how did your broker know you wanted to retire in ten years?" The answer is, "He asked!" These questions might include:

   a. Who is the client? This addresses who the system is really being built to satisfy.

   b. What is a highly successful solution really worth to this client? This addresses the relative value of the solution.

   c. What is the real reason for this request? This might lead to alternative approaches that are more satisfactory to the customer.

   d. Who should be on the development team? This may identify particular skills or knowledge that the customer recognizes as necessary.

e. What are the customer's trade-offs between time and value? This gives a better perspective on urgency.

4. Ask questions that will help you understand the product better. These might include:

   a. What problem does this product solve?

   b. What problem could this product create? It is a good idea to think about this early. This might also help to bound the problem area.

   c. What environment is this product likely to encounter? Again, this helps to bound the request, and it may help identify some of the restrictions on the solution.

   d. What kind of precision is required or desired in the product?

Scenarios. A scenario is a form of prototyping used to describe a customer request. In a scenario, examples of how a solution might react are examined. Again, this is a very natural process, but one that needs to be prepared to work effectively. Scenarios are built by determining how implementation might satisfy the request. Assumptions, problems, and the customers' reactions to the scenarios are used for further evaluation.

Once the interviews and scenarios are progressing, a work statement can be formed. A work statement is a textual description of the customer's interpretation of the request. This may range from very formal to very abstract. The purpose of interviews and scenarios is to clarify the work statement.

Ambiguity Resolution. Ambiguity resolution is an attempt to identify all the possible misinterpretable aspects of a work statement. Examples of ambiguous areas are: terms, the definitions that are often interpreted differently; acronyms that are not clearly defined; and misleading statements of what is required. Question everything; nothing should be assumed while reading the work statement.

The output of this step is a redefined work statement, possibly with a list of definitions. The list of definitions should be the starting point for a future data dictionary.

Context Analysis. Context analysis is an early modeling of the system to identify key objects as named links. This helps to organize pieces of the request, and provides connectivity to highlight communication.

Clarification and Prioritization. The final step in requirements elicitation is to create a list of the desired functions, attributes, constraints, preferences, and expectations. This moves the work statement towards a more formal software requirements specification:

1. Functions are the "what" of the product. They describe what the product must do or accomplish to satisfy the customer's request. Functions should be classified as *evident, hidden,* or *frill.* Once this classification is complete, the developer has set the stage for what must be demonstrated to the customer—the *evident* functions. and what must be developed to satisfy the customer—the *hidden* functions. The *frills* are kept on a standby list to be obtained, if possible.

2. Attributes are characteristics of the product. They describe the way the product should satisfy customer needs. Attributes and attribute lists (linear descriptions of how the attribute is recognized) can be classified as *must, want,* and *ignore.* Attributes and attribute lists are attached to the functions described above. To satisfy the

customer all *must* attributes have to be met, and as many *want* attributes as possible should be met. *Ignore* attributes are ignored.

3. Constraints are mandatory conditions placed on an attribute. These may come from regulations, policies, or customer experience. (Developers should always attempt to recognize and validate all constraints. Constraints limit the options available and they should be investigated before they limit the system.)

4. Preferences are desirable, but optional, conditions placed on an attribute. Preferences help to satisfy the customer, so they should be achieved whenever possible.

5. Expectations are the hardest items to capture. These are what the customer expects from the product; they are seldom clearly defined. Every attempt should be made to determine what the customer expects the system to do, and these expectations should be met if possible. If the customer's expectations are exorbitant, then these issues should be addressed at the project onset to avoid later disappointment.

The output of the requirements elicitation process is a well-formed, redefined work statement clearly stating the customer's request. This work statement provides the basis for the requirements analysis.

Requirements Analysis. Requirements analysis involves taking the work statement and creating a model of the customer's request. The type of request should determine which technique to select.

Requirements Specification. This specification, called the SRS, is the formal definition of the customer's problem. The requirements specification combines the knowledge gained during requirements elicitation (attributes, constraints, and preferences) with the model produced during requirements analysis.

Characteristics of a Good SRS. This section should describe the characteristics of a good SRS. The following are offered as criteria:

Correct. "An SRS is correct if and only if every requirement stated therein represents something required of the system to be built." [75]

Nonambiguous. "An SRS is nonambiguous if and only if every requirement stated therein has only one interpretation." [75]

Complete. "An SRS is complete if it possesses the following four qualities:

1. Everything that the software is supposed to do is included in the SRS.
2. All definitions of software responses to all realizable classes of input data, in all realizable classes of situations.
3. All pages are numbered; all figures and tables are numbered, named, and referenced; all terms and units of measure are provided; and all referenced material and sections are present.
4. No sections are marked "To Be Determined (TBD)." [75]

Verifiable. An SRS is verifiable only if every requirement stated therein is verifiable. A requirement is verifiable if there exists some finite cost effective process with which a person, or machine, can check that the actual as-built software product meets the requirement.

Consistent. An SRS is consistent, if and only if, no subset of individual requirements stated therein conflict. This may be manifested in a number of ways:
  1. Conflicting terms
  2. Conflicting characteristics
  3. Temporal inconsistency*

Modifiable. "An SRS is modifiable if its structure and style are such that any necessary changes to the requirements can be made easily, completely, and consistently."[75]

Traceable. "An SRS is traceable if the origin of each of its requirements is clear and if it facilitates the referencing of each requirement in future development or enhancement documentation." [75]

Usable during Operation and Maintenance. This is a combination of being modifiable, as described above, and being understandable to users of the SRS. The SRS should last as long as the system built from it and always accurately reflect the state of the system. SRS requirements that do not meet the above criteria should be considered a risk or a reportable issue for the project manager's open issues list.

Standards and Practices. This section should provide references to applicable corporate policies and procedures, and external standards if appropriate. The following is provided as an example:

> [PROJECT ABBREVIATION] Projects will document and maintain the current status of all customer approved SRSs. At this time, the standard for software requirements management is limited to the need to establish and follow the format of Appendix A to create specifications for each new project. This format will be followed unless the program is required, by contract, to use another format. As [Company Name]'s maturity within the Capability Maturity Model increases, more detailed standards and practices will evolve. These standards will be based on metrics to ensure consistent capability to complete the requirements.
>
> In developing the software requirements specification, the program manager will incorporate all existing [Company Name] standards and policies for software, including, but not limited to software configuration management and software quality assurance.

Software Measurement and Metrics. This section should describe all measurement activities associated with an organization's management of software requirements. The following is provided as an example:

> Software measurement and metrics will be applied to the process of software requirements management on a continual basis in order to ensure that customer requirements are met and to improve the SRM process. Measurement from a [PROJECT ABBREVIATION] perspective means measuring progress towards the [PROJECT ABBREVIATION] goal of accurately identifying project software requirements. (Reference project level software measurement and metrics plan if appropriate.)
>
> In order to improve, the results of projects must be gathered and include deviations from the SRSs. Recording and learning from the lessons of early projects will improve the software requirements management process and serve to move [PROJECT ABBREVIATION] toward the goal of effective requirements management. Both corporate and project configuration management will help review SRSs and identify deviations.

---

*Temporal inconsistency is when two or more parts of the SRS require the product to obey contradictory timing constraints.

Verification and Validation. This section should describe the minimum acceptable verification and validation of software requirements. Verification and validation are two important aspects of any project software life cycle. They provide the checks and balances to the development process that ensures that the system is being built (validation). According to IEEE/EIA 12207.0, verification is "Confirmation by examination and provision of objective evidence that specified requirements have been fullfilled" and validation is "Confirmation by examination and provision of objective evidence that the particular requirements for a specific intended use are fullfilled."[2] In other words, Verification is "Are we building the right product?" and Validation is "Are we building the product right?"

Verification and validation differ in scope and purpose. Verification is concerned with comparing the results of each activity of the software life cycle with the specification from the previous activity. The scope of verification, therefore, is limited to two activities (current and prior). For this reason, the context of verification changes depending upon the pair of activities under review.

Validation takes a bird's-eye view of requirements and the implementation of those requirements. Each functional and nonfunctional requirement listed in the SRS will be testable by one of four techniques. The SRS will reference a requirements traceability matrix (RTM), that lists all requirements and the method used to validate that requirement. Four possible methods of validation are:

1. **Demonstration.** The end product is executed for the customer to demonstrate that the desired properties have been met.
2. **Test.** The end product is executed through predefined, scripted tests for the customer to show functionality and demonstrate correctness.
3. **Analysis.** Results of the product execution are examined to ensure program correctness.
4. **Inspection.** The product itself is inspected to demonstrate that it is correct.

Software Configuration Management. This section should provide a cross-reference to the supporting software configuration management plan and associated activities. The following is provided as an example:

> The software project management plan will present a schedule for developing the SRS. The schedule will indicate the expected time for baselining the completed SRS. The baselined SRS will be placed under configuration management defined in the program configuration management process.

Developing a Software Requirements Specification. This section should provide detailed guidance in support of the development of a SRS. The following is provided as guidance:

> Each program manager should ensure all new projects develop an SRS and submit through the Corporate SQA Manager to [COMPANY NAME] Management for approval. Within each of the sections listed in Appendix A, a general explanation of the section contents is given. Appendix A should be considered a boilerplate for developing your SRS. If a section does not pertain to your project, the following should appear below the section heading "This section is not applicable to this plan," together with appropriate reasons for the exclusion. Additional sections and appendices may be added as needed. Some of the re-

quired material may appear in other project documents. If so, then references to these documents should be made in the body of the software requirements specification. In any case, the contents of each section of the plan should be specified either directly or by reference to another document.

Appendix A. Project Software Requirements Specification. This appendix should contain a document template for project use.

Appendix B. Template for Requirements Traceability Matrix. This appendix should contain a traceability matrix template for project use.

PROJECT PLANNING

Project planning lays the foundation that can determine the success or failure of a software project. It is crucial that appropriate attention be given to project planning to enable project managers to derive useful information for making decisions to achieve schedule and budget goals. The CMMI®-SW elevates risk management to a process area, with its own set of required goals and practices as they apply to software. In specifies that as specific project activities, risk management efforts must be institutionalized through written organizational policies, plans, allocated resources, assigned responsibilities, training, configuration management of work products, quality audits, and management reviews. The Level 2 project management process areas are Project Planning, Project Monitoring and Control, and Supplier Agreement Management.

IEEE Std 12207.0, Standard for Information Technology—Software Life Cycle Processes [39], describes 17 processes spanning the entire life cycle of a software product or service. Even if an organization's processes were defined using other sources, the standard is useful in characterizing the essential characteristics of these software processes and should be considered prior to the implementation of process improvement activities. Referencing this standard and reviewing what is required for each of these primary process areas can provide additional guidance in support of the activities associated with project planning. It is important that IEEE 12207 be considered prior to the implementation of any process improvement activities associated with project planning.

CMMI®-SW Goals

The CMMI®-SW Level 2 is focused on establishing estimates, plan development, and plan commitment in support of project activities. The CMMI®-SW places emphasis on having a detailed project work breakdown structure (WBS). The planning, and maintenance, of project data items and their contents has also been added to the list of project management concerns. The CMMI®-SW specifically requires the execution the project according to estimations. Estimation focuses on size and complexity while effort and cost, and schedule are determined and established respectively based on this estimate. The CMMI®-SW also addresses the reconciliation (revision) of the software project management plan to reflect changes in available and estimated resources.

SG 1. Establish Estimates
Estimates of project planning parameters are established and maintained.

A detailed WBS must be included in, or referenced by, the software project management plan. IEEE Std 1058-1998, IEEE Standard for Software Project Management Plans [15], requires that all roles, responsibilities, authorities, schedule, budgets, resource requirements, risk factors, and work products for each supporting process be identified. IEEE Std 1058 suggests that all aspects of project organization be reported, as well as a description of all technical, managerial, and supporting processes. It also requires that a section of the software project management plan specify the cost and schedule as well as the methods, tools, and techniques used to estimate project cost, schedule, and resource requirements.

SG 2. Develop a Project Plan
A project plan is established and maintained as the basis for managing the project.

IEEE Standard 1058 provides a description of the elements critical to a software project management plan. This standard provides both the outline and suggested content for a project plan. Suggested document maintenance activities are also addressed. The CMMI® has a requirement to establish and maintain a plan that describes the performance of the project planning process. IEEE Standard 1058 provides support for this process area (PA) and can be used to support the development of a plan that provides the detailed description of this process.

SG 3. Obtain Commitment to the Plan
Commitments to the project plan are established and maintained.

Obtaining and managing all commitments to the software project management plan are detailed in Section 4.5.3.5—Reporting Plan, and Section 4.5.4—Risk Management Plan of IEEE Std 1058.

GG 2. Institutionalize a Managed Process
The process is institutionalized as a managed process.

IEEE Std 1058-1998 is an effective instrument for the implementation of Level 2 PAs. However, information regarding the measurement and metrics required in support of software project planning activities needs to be added and stated explicitly. It is also important to note that while the various Level 2 PAs are addressed by the major headings of IEEE 1058, the details required to support the CMMI® can be lost if each section is not carefully addressed while bearing the specific CMMI® project planning commitments, abilities, measurements, verification, and activities in mind.

It is important to note that IEEE Std 1219-1998, IEEE Standard for Software Maintenance [22], provides detailed guidance in support of requirements, planning, management, execution, and documentation of software maintenance activities. The basic process model described supports the input, process, output, and control process model for software maintenance. Information is provided regarding the types of measurement and metrics used in support of software maintenance activities with no predisposition of lifecycle model (e.g., incremental, spiral). This standard also provides specific guidelines for maintenance plan development. Table 5-3 describes the specific and generic goals and practices, typical work products, and commonly associated plans or artifacts in support of the PA.

Table 5-3. Project planning goals and practices [61]

Specific and generic goals/practices and typical work products	Plan or artifact
SG1. Establish Estimates	
SP1.1. Estimate the scope of the project	
WBS, task descriptions, and work package descriptions	Software project management plan
SP1.2. Establish estimates of work product and task attributes	Software project management plan
Technical approach	
Size and complexity of tasks and work products	Software project management plan
Estimating models	Software project management plan
Attribute estimates	Software project management plan
SP1.3. Define product lifecycle	
Project lifecycle phases	Software project management plan
SP1.4. Determine estimates of effort and cost	
Estimation rationale	Software project management plan
Project effort estimates	Software project management plan
Project cost estimates	Software project management plan
SG2. Develop a Project Plan	
SP2.1. Establish budget and schedule	
Project schedules	Software project management plan
Schedule dependencies	Software project management plan
Project budget	Software project management plan
SP2.2. Identify project risks	
Identified risks	Project plan
Risk impacts and probability of occurrence	Software project management plan
Risk priorities	Software project management plan
SP2.3. Plan for data management	
Data management plan	Configuration management plan
Master list of managed/collected data	Configuration management plan
Data content and format description	Configuration management plan
Data requirements lists for acquirers and for suppliers	Configuration management plan
Privacy and security requirements	Configuration management plan
Security procedures	Configuration management plan
Mechanism for data retrieval, reproduction, and distribution	Configuration management plan
Schedule for collection of project data	Software project management plan
SP2.4. Plan for project resources	
WBS work packages	Software project management plan, postdevelopment phase
WBS task dictionary	Software project management plan
Staffing requirements	Software project management plan
Critical facilities/equipment list	Software project management plan
Process/workflow definitions and diagrams	Software project management plan
Program administration requirements list	Software project management plan
SP2.5. Plan for needed knowledge and skills	
Inventory of skill needs	Software project management plan
Staffing and new hire plans	Software project management plan
Databases (e.g., skills and training)	Software project management plan

Table 5-3. *Continued*

Specific and generic goals/practices and typical work products	Plan or artifact
SG2. Develop a Project Plan (*cont.*)	
SP2.6. Plan stakeholder involvement	
Stakeholder involvement plan	Software project management plan
SP2.7. Establish the project plan	
Overall project plan	Software project management plan, postdevelopment activities
SG3. Obtain Commitment to the Plan	
SP3.1. Review plans that affect the project	
Records of project plan reviews	Software project management plan, software project plan walk through checklist
SP3.2. Reconcile work and resource levels	
Revised methods and corresponding estimating parameters	Software project management plan
Renegotiated budgets	Software project management plan
Revised schedules	Software project management plan
Revised requirements list	Software requirements specification or database
Renegotiated stakeholder agreements	Software project management plan
SP3.3. Obtain plan commitment	Documented commitments and requests for commitments
GG2. Institutionalize a Managed Process	
GP2.1. Establish an organizational policy	Organizational policy supporting project planning
GP2.2. Plan the process	Software project management plan including data management and stakeholder involvement planning
GP2.3. Provide resources	
Identification of resources used in support of planning	Software project management plan
GP2.4. Assign responsibility	Software project management plan or organizational policy
GP2.5. Train people	Software project management plan or training plan
GP 2.6. Manage configurations	Configuration management plan
GP 2.7. Identify relevant stakeholders	Supporting artifacts could include CCB charter and software project management plan
GP 2.8. Monitor and control the process	
How planning volatility will be reported	Description may be included in software metrics and measurement plan or software project management plan

(*continued*)

Table 5-3. *Continued*

Specific and generic goals/practices and typical work products	Plan or artifact
GG2. Institutionalize a Managed Process (*cont.*)	
GP 2.9. Objectively evaluate adherence	Reviews are identified in software quality assurance plan
GP 2.10. Review status with higher-level management	Reviews and reporting are identified in software project management plan
GG3. Institutionalize a Defined Process	
GP 3.1. Establish a defined process	Software project management plan
GP 3.2. Collect improvement information	Requirements management plan, software metrics and measurement plan, or software project management plan

Software Project Management Plan

IEEE std 1058-1998, IEEE Standard for Software Project Management Plans [15], and IEEE/EIA 12207.0, IEEE Standard for Life Cycle Processes [39], are effective instruments for the implementation of Level 2 PAs. However, information regarding the measurement and metrics required in support of software project planning activities needs to be added and stated explicitly. It is also important to note that while the various Level 2 PAs are addressed by the major headings of IEEE 1058, the details required to support the CMM® can be lost if each section is not carefully addressed while bearing the specific CMM® project planning commitments, abilities, measurements, verification, and activities in mind.

The modification of the recommended Software Project Management Plan (SPMP) table of contents to support the goals of the CMMI® more directly is shown in Table 5-4.

Software Project Management Plan Document Guidance

The following provides section-by-section guidance in support of the creation of a SPMP. The SPMP should be considered to be a living document. The SPMP, and in particular, any associated schedules, should change to reflect any required change during the life cycle of a project. This guidance should be used to help define a management process and should reflect the actual processes and procedures of the implementing organization. Additional information is provided in the document template, *Software Project Management Plan.doc,* located on the CD-ROM accompanying this book. Additional information is provided in Appendix C, Software Process Work Products, that describes the work breakdown structure, workflow diagram, and stakeholder involvement matrix work products.

Project Overview. This paragraph should briefly state the purpose, scope, and objectives of the system and the software to which this document applies. It should describe the general nature of the system and software; summarize the history of system development, operation, and maintenance; identify the project sponsor, acquirer, user, developer, and support agencies; and identify current and planned operating sites. The project overview

Table 5-4. Software project management plan document outline

Title Page
Revision Page
Table of Contents
1. Introduction
   1.1 Project Overview
   1.2 Project Deliverables
   1.3 Document Overview
   1.4 Acronyms and Definitions
2. References
3. Project Organization
   3.1 Organizational Policies
   3.2 Process Model
   3.3 Organizational Structure
   3.4 Organizational Boundaries and Interfaces
   3.5 Project Responsibilities
4. Managerial Process
   4.1 Management Objectives and Priorities
   4.2 Assumptions, Dependencies, and Constraints
   4.3 Risk Management
   4.4 Monitoring and Controlling Mechanisms
   4.5 Staffing Plan
5. Technical Process
   5.1 Tools, Techniques, and Methods
   5.2 Software Documentation
   5.3 Project Support Functions
6. Work Packages
   6.1 Work Packages
   6.2 Dependencies
   6.3 Resource Requirements
   6.4 Budget and Resource Allocation
   6.5 Schedule
7. Additional Components

should also describe the relationship of this project to other projects, as appropriate, addressing any assumptions and constraints. This section should also provide a brief schedule and budget summary. This overview should not be construed as an official statement of product requirements. Reference to the official statement of product requirements should be provided in this subsection of the SPMP.

Project Deliverables. This subsection of the SPMP should list all of the items to be delivered to the customer, the delivery dates, delivery locations, and quantities required to satisfy the terms of the project agreement. This list of project deliverables should not be construed as an official statement of project requirements.

Document Overview. This subsection should summarize the purpose and contents of this document and describe any security or privacy considerations that should be considered associated with its use. This subsection of the SPMP should also specify the plans for producing both scheduled and unscheduled updates to the SPMP. Methods of disseminating the updates should be specified. This subsection should also specify the mechanisms used to place the initial version of the SPMP under change control and to control subsequent changes to the SPMP.

Acronyms and Definitions. This subsection should identify acronyms and definitions used within the project SPMP. The project SPMP should only list acronyms and definitions used within the SPMP.

References. This section should identify the specific references used within the project SPMP. The project SPMP should only contain references used within the SPMP.

Organizational Policies. This subsection of the SPMP should identify all organizational policies relative to the software project.

Process Model. This subsection of the SPMP should specify the (life-cycle) software development process model for the project, describe the project organizational structure, identify organizational boundaries and interfaces, and define individual or stakeholder responsibilities for the various software development elements.

Organizational Structure. This subsection should describe the makeup of the team to be used for the project. All project roles and stakeholders should be identified and a description of the internal management structure of the project should be provided. Diagrams may be used to depict the lines of authority, responsibility, and communication within the project.

Organizational Boundaries and Interfaces. This subsection should describe the limits of the project, including any interfaces with other projects or programs, the application of the program's SCM and SQA (including any divergence from those plans), and the interface with the project's customer. It should describe the administrative and managerial boundaries between the project and each of the following entities: the parent organization, the customer organization, subcontracted organizations, or any other organizational entities that interact with the project. In addition, the administrative and managerial interfaces of the project support functions, such as configuration management, quality assurance, and verification, should be specified in this subsection.

Project Responsibilities. This subsection should describe the project's approach through a description of the tasks required to complete the project (e.g., requirements → design → implementation → test) and any efforts (update documentation, etc.) required to successfully complete the project. It should state the nature of each major project function and activity, and identify the individuals or stakeholders who are responsible for those functions and activities.

Managerial Process. This section should specify management objectives and priorities; project assumptions, dependencies, and constraints; risk management techniques; monitoring and controlling mechanisms required; and the staffing plan.

Management Objectives and Priorities. This subsection should describe the philosophy, goals, and priorities for management activities during the project. Topics to be specified may include, but are not limited to, the frequency and mechanisms of reporting to be used; the relative priorities among requirements, schedule, and budget for this project; risk management procedures to be followed; and a statement of intent to acquire, modify, or use existing software.

Assumptions, Dependencies, and Constraints. This subsection should state the assumptions on which the project is based, the external events the project is dependent upon, and the constraints under which the project is to be conducted.

Risk Management. This subsection should identify the risks for the project. Completed risk management forms should be maintained and tracked by the project leader with associated project information. These forms should be reviewed at weekly staff meetings. Risk factors that should be considered include contractual risks, technological risks, risks due to size and complexity of the project, risks in personnel acquisition and retention, and risks in achieving customer acceptance of the product.

Monitoring and Controlling Mechanisms. This subsection should define the reporting mechanisms, report formats, information flows, review and audit mechanisms, and other tools and techniques to be used in monitoring and controlling adherence to the SPMP. A typical set of software reviews is listed in Appendix C. Project monitoring should occur at the level of work packages. The relationship of monitoring and controlling mechanisms to the project support functions should be delineated in this subsection. This subsection should also describe the approach to be followed for providing the acquirer or its authorized representative access to developer and subcontractor facilities for review of software products and activities.

Staffing Plan. This subsection should specify the numbers and types of personnel required to conduct the project. Required skill levels, start times, duration of need, and methods for obtaining, training, retaining, and phasing out of personnel should be specified.

Technical Process. This section should specify the technical methods, tools, and techniques to be used on the project. In addition, the plan for software documentation should be specified, and plans for project support functions such as quality assurance, configuration management, and verification and validation may be specified.

Tools, Techniques, and Methods. This subsection of the SPMP should specify the computing system(s), development methodology(s), team structures(s), programming language(s), and other notations, tools, techniques, and methods to be used to specify, design, build, test, integrate, document, deliver, modify, or maintain (or both, as appropriate) the project deliverables.

This subsection should also describe any tools (compilers, CASE tools, and project management tools), any techniques (review, walk-through, inspection, and prototyping) and the methods (object-oriented design, rapid prototyping) to be used during the project.

Software Documentation. This subsection should contain, either directly or by reference, the documentation plan for the software project. The documentation plan should specify the documentation requirements, and the milestones, baselines, reviews, and sign-offs for software documentation. The documentation plan may also contain a style guide, naming conventions, and documentation formats. The documentation plan should provide a summary of the schedule and resource requirements for the documentation effort. IEEE Std for Software Test Documentation (IEEE Std 829-1998) [5] provides a standard for software test documentation.

Project Support Functions. This subsection should contain, either directly or by reference, plans for the supporting functions for the software project. These functions may include, but are not limited to, configuration management, software quality assurance, and verification and validation.

Work Packages. This section of the SPMP should specify the work packages, identify the dependency relationships among them, state the resource requirements, provide the allocation of budget and resources to work packages, and establish a project schedule. The work packages for the activities and tasks that must be completed in order to satisfy the project agreement must be described in this section. Each work package should be uniquely identified; identification may be based on a numbering scheme and descriptive titles. A diagram depicting the breakdown of activities into subactivities and tasks may be used to depict hierarchical relationships among work packages.

Dependencies. This subsection should specify the ordering relations among work packages to account for interdependencies among them and dependencies on external events. Techniques such as dependency lists, activity networks, and the critical path may be used to depict dependencies.

Resource Requirements. This subsection should provide, as a function of time, estimates of the total resources required to complete the project. Numbers and types of personnel, computer time, support software, computer hardware, office and laboratory facilities, travel, and maintenance requirements for the project resources are typical resources that should be specified.

Budget and Resource Allocation. This subsection should specify the allocation of budget and resources to the various project functions, activities, and tasks. Defined resources should be tracked.

Schedule. This subsection should be used to capture the project's schedule, including all milestones and critical paths. Options include Gantt charts (Milestones Etc.™ or Microsoft Project™), Pert charts, or simple time lines.

Additional Components. This section should address additional items of importance on any particular project. This may include subcontractor management plans, security plans, independent verification and validation plans, training plans, hardware procurement plans, facilities plans, installation plans, data conversion plans, system transition plans, or product maintenance plans.

Data Management

Data management planning is functionally described, rather than defined, in the CMMI®. So the question arises as to where to address it in the project documentation. The planning for data management may be addressed in the software project management plan or explicitly addressed in a separate data management plan. Data management and configuration management activities are closely related and, as such, data management controls and configuration may be addressed in the configuration management plan. Configuration management should include data management functions. All data associated with a project must be managed (i.e., put under configuration management). Larger projects often have a data manager who is responsible for maintaining the integrity of the problem reporting systems, documentation, and so on. Configuration management identifies, controls, traces the changes, reports status, and audits. Data management supports these activities.

Configuration management is required to control all product data (i.e., requirements, specifications, design, development code, etc.). All data that are necessary to create, reuse, modify, and support a product fall under the responsibility of configuration management (CM). Data management is a tool that CM uses to accomplish these tasks and, as such, is a subset of CM.

The data collection activities, if not provided in the project plan, may be detailed in any associated software measurement and metrics plan. Data management requirements are interwoven throughout the CMMI and these requirements should be kept in mind while developing all other project or organizational documentation.

Stakeholder Involvement

The CMMI requires that all stakeholder involvement be monitored against the project plan. IEEE Std 1058 addresses the identification of all key participants (project stakeholders) associated with a development effort. This standard requires the definition of the organizational structure, all customer relationships, roles, responsibilities, resources, and how involvement is monitored. The CMMI has placed special emphasis on stakeholder involvement in the software development process and, as such, should be specifically addressed in the software project management plan. Refer to Appendix C for an example of a stakeholder involvement matrix.

PROJECT MONITORING AND CONTROL

The CMMI®-SW has elevated the monitoring of project commitments to the specific practice level and strongly emphasizes the effective monitoring of risk and stakeholder involvement. The project management plan becomes the basis for monitoring activities, communicating status, and taking corrective action. Progress is measured against this plan by comparing the actual work product and task attributes effort, cost, and schedule at prescribed milestones within the project schedule or work breakdown structure. Technical reviews, management reviews, and audits are the key project monitoring and control activities; they are discussed in the following material. Additional information is provided in Appendix C that describes the Open Issues List work product.

IEEE Std 12207.0, Standard for Information Technology—Software Life Cycle Processes [39], describes 17 processes spanning the entire life cycle of a software product or service. Even if an organization's processes were defined using other sources, the standard is useful in characterizing the essential characteristics of these software processes and should be considered prior to the implementation of process improvement activities.

Referencing this standard and reviewing what is required for each of these primary process areas can provide additional guidance in support of the activities associated with project monitoring and control. It is important that IEEE 12207 be considered prior to the implementation of any process improvement activities associated with project monitoring and control.

CMMI®-SW Goals

The focus of this PA is on the processes supporting the effective management of a software project. This begins with the documentation of the project plan and provides for the required processes supporting the identification and appropriate handling of deviations from the planning documentation. It addresses the effective management of any required corrective action when project performance deviates significantly from projected baselines.

Progress is primarily determined by comparing the actual work product and task attributes effort, cost, and schedule to the planned estimates. These comparisons occur at prescribed milestones that are identified in the project schedule or work breakdown structure. This visibility enables a proactive response when performance deviates significantly from what is planned. A deviation is significant if, when left unresolved, it precludes the project from meeting its objectives.

SG1. Monitor Project Against Plan
Actual performance and progress of the project are monitored against the project plan.

The IEEE 1028, IEEE Standard for Software Reviews [12], and IEEE 12207.0, IEEE Standard for Life Cycle Processes [39], describe a set of review types and processes applicable throughout the software life cycle. These standards describe technical reviews, inspections, and walkthroughs. The necessary steps required for each review type are described in detail. These standards also describe the software audit process. IEEE Std 1028 also provides an annex that supports the process of review selection, providing a comparison of the different types of reviews.

SG2. Manage Corrective Action to Closure
Corrective actions are managed to closure when the project's performance or results deviate significantly from the plan.

Many organizations use informal reviews, not implementing the reviews formally as an integrated, required, part of their software process. Too often, review types are mixed and simply named "reviews" or "inspections." A walkthrough, focused on achieving consensus, is something very different from an inspection, which is focused on formal defect finding. It is important to have a clear understanding of the desired benefit from the implementation of each type of review prior to its implementation. IEEE Std 1028 provides clear definitions and straightforward process steps. This standard describes the different review types and their specific objectives, providing a complete reference framework and common terminology. Reviewing, when applied correctly and early in the process, remains the most effective and efficient defect detection technique [102].

GG2. Institutionalize a Managed Process
The process is institutionalized as a managed process.

The CMMI®-SW specifically requires that the process be institutionalized as a managed process and that there be a plan for performing project monitoring and control activities. The IEEE standards listed in the previous matrix provide support for these requirements. Table 5-5 describes the specific and generic goals and practices, typical work products, and commonly associated plans or artifacts in support of the PA.

Technical Reviews

The information provided here is based upon the recommendations provided in IEEE Std 1028-1997, IEEE Standard for Software Reviews. Technical reviews are an effective way to detect and identify problems early in the software development process. The criteria for technical reviews are presented here to help organizations define their review practices.

Introduction. Technical reviews are an effective way to evaluate a software product by a team of qualified personnel to identify discrepancies from specifications and standards. Whereas technical reviews identify anomalies, they may also provide the recommendation and examination of various alternatives. The examination need not address all aspects of the product and may only focus on a selected aspect of a software product.

Software requirements specifications, design descriptions, test and user documentation, installation and maintenance procedures, and build processes are all examples of items subject to technical reviews.

Responsibilities. The roles in support of a technical review are:

Decision maker. The decision maker is the individual requesting the review and determines whether objectives have been met.

Review leader. The review leader is responsible for the review and must perform all administrative tasks (including summary reporting) to ensure that the review is conducted in an orderly manner and meets its objectives.

Recorder. The recorder is responsible for the documentation of all anomalies, action items, decisions, and recommendations made by the review team.

Technical staff. The technical staff should actively participate in the review and evaluation of the software product.

The following roles are optional and may also be established for the technical review:

Management staff. Management staff may participate in the technical review for the purpose of identifying issues that will require resolution by management.

Customer or user representative. The role of the customer or user representative should be determined by the review leader prior to the review.

Input. Input to the technical review should include a statement of objectives, the software product being examined, existing anomalies and review reports, review procedures, and any standard against which the product is to be examined. Anomaly categories should be defined and available during the technical review. For additional information in support of the categorization of software product anomalies, refer to IEEE Std 1044 [13].

Table 5-5. Project monitoring and control goals and practices [61]

Specific and generic goals/practices and typical work products	Plan or artifact
SG1. Monitor Project Against Plan	
SP1.1. Monitor project planning parameters	
Records of project performance	Software project management plan
Records of significant deviations	Reflected changes in schedule
SP1.2. Monitor commitments	Software project management plan
Records of commitment reviews	
SP1.3. Monitor project risks	
Records of project risk monitoring	Risk management plan or software project management plan
SP1.4. Monitor data management	
Records of data management	Configuration management plan or software project management plan
SP1.5. Monitor stakeholder involvement	
Records of stakeholder involvement	Software project management plan, CCB charter, teaming agreement
SP1.6. Conduct progress reviews	
Documented project review results	May be informal or software project management plan
SP1.7. Conduct milestone reviews	
Documented review results	Software project management plan or software quality assurance plan
SG2. Manage Corrective Action to Closure	
SP 2.1. Analyze issues	
List of issues needing corrective action	Software project management plan
SP 2.2. Take corrective action	
Corrective action plan	Software project management plan
SP 2.3. Manage corrective action	
Corrective action results	Software project management plan
GG2. Institutionalize a Managed Process	
GP 2.1. Establish an organizational policy	Organizational policy supporting project monitoring and control
GP 2.2. Plan the process	Software project management and/or software measurement and metrics plan
GP 2.3. Provide resources	
Identification of resources used in support of project monitoring and control	Software project management plan
GP 2.4. Assign responsibility	Software project management plan or organizational policy
GP 2.5. Train people	Software project management plan or training plan

Table 5-5. *Continued*

Specific and generic goals/practices and typical work products	Plan or artifact
GG2. Institutionalize a Managed Process (*cont.*) *GP 2.6. Manage configurations* Description of how work products and all associated artifacts in support of project monitoring and control are placed under configuration management	Configuration management plan
GP 2.7. Identify relevant stakeholders	CCB charter and software project management plan
GP 2.8. Monitor and control the process How project volatility will be reported	Software measurement and metrics plan, requirements management plan, unit test plan, or software project management plan
GP 2.9. Objectively evaluate adherence	Software quality assurance plan
GP 2.10. Review status with higher-level management	Software project management plan and software quality assurance plan
GG3. Institutionalize a Defined Process *GP 3.1. Establish a defined process* A description of the project monitoring and control process	Software project management plan
GP 3.2. Collect improvement information	Requirements management plan, measurement and metrics plan, or software project management plan

Authorization. All technical reviews should be defined in the SPMP. The plan should describe the review schedule and all allocation of resources. In addition to those technical reviews required by the SPMP, other technical reviews may be scheduled.

Preconditions. A technical review shall be conducted only when the objectives have been established and the required review inputs are available.

Procedures. These consist of:

Management Preparation. Managers are responsible for ensuring that all reviews are planned, that all team members are trained and knowledgeable, and that adequate resources are provided. They are also responsible for ensuring that all review procedures are followed.

Planning the Review. The review leader is responsible for the identification of the review team and their assignment of responsibility. The leader should schedule and announce the meeting, and prepare participants for the review by providing them with the required material and collecting all comments.

Overview of Review Procedures. The team should be presented with an overview of the review procedures. This overview may occur as a part of the review meeting or as a separate meeting.

Overview of the Software Product. The team should receive an overview of the software product. This overview may occur either as a part of the review meeting or as a separate meeting.

Preparation. All team members are responsible for reviewing the product prior to the review meeting. All anomalies identified during this prereview process should be presented to the review leader. Prior to the review meeting, the leader should classify all anomalies and forward these to the author of the software product for disposition.

The review leader is also responsible for the collection of all individual preparation times, which are used to determine the total preparation time associated with the review.

Examination. The review meeting should have a defined agenda that should be based upon the premeeting anomaly summary. Based upon the information presented, the team should determine whether the product is suitable for its intended use, conforms to appropriate standards, and is ready for the next project activity. All anomalies should be identified and documented.

Rework/Follow-up. The review leader is responsible for verifying that the action items assigned in the meeting are closed.

Exit Criteria. A technical review shall be considered complete when all follow-up activities have been completed and the review report has been published.

Output. The output from the technical review should consist of the project being reviewed, a list of the review team members, a description of the review objectives, and a list of resolved and unresolved software product anomalies. The output should also include a list of management issues, all action items and their status, and any recommendations made by the review team.

Management Reviews

The information provided here is based upon the recommendations provided in IEEE Std 1028-1997, IEEE Standard for Software Reviews. Management reviews are an effective way to detect and identify problems early in the software development process. The criteria for management reviews are presented here to help organizations define their review practices. It is important to remember that management reviews are not only about the specific review of products, but also may cover aspects of the software process.

Introduction. The purpose of a management review is to monitor project progress and to determine the status according to documented plans and schedules. Reviews can also be used to evaluate the effectiveness of the management approaches. The effective use of management reviews can support decisions about corrective action, resource allocation, or changes in scope.

Management reviews may not address all aspects of a given project during a single review but may require several review cycles to completely evaluate a software product. Examples of software products subject to management review include:

a. Anomaly reports
b. Audit reports
c. Back-up and recovery plans
d. Contingency plans
e. Customer complaints
f. Disaster plans
g. Hardware performance plans
h. Installation plans
i. Maintenance plans
j. Procurement and contracting methods
k. Progress reports
l. Risk management plans
m. Software configuration management plans
n. Software project management plans
o. Software quality assurance plans
p. Software safety plans
q. Software verification and validation plans
r. Technical review reports
s. Software product analyses
t. Verification and validation reports

Responsibilities. The management personnel associated with a given project should carry out management reviews. The individuals qualified to evaluate the software product should perform all management reviews. These individuals are:

Decision maker. The review is conducted for the decision maker and it is up to this individual to determine whether the objectives of the review have been met.

Review leader. The review leader is responsible for all administrative tasks required in support of the review. The review leader should ensure that all planning and preparation have been completed, that all objectives are established and met, and that the review is conducted and review outputs are published.

Recorder. The recorder is responsible for the documentation of all anomalies, action items, decisions, and recommendations made by the review team.

Management staff. The management staff is responsible for carrying out the review. They are responsible for active participation in the review.

Technical staff. The technical staff are responsible for providing the information required in support of the review.

Customer or user representative. The role of the customer or user representative should be determined by the review leader prior to the review.

Input. Input to the management review should include the following:

a. A statement of review objectives
b. The software product being evaluated
c. The software project management plan
d. Project status
e. A current anomalies list
f. Documented review procedures
g. Status of resources, including finance, as appropriate
h. Relevant review reports
i. Any associated standards
j. Anomaly categories

Authorization. The requirement for conducting management reviews should initially be established in the project planning documents. The completion of a software product or completion of a scheduled activity may initiate a management review. In addition to those management reviews required by a specific plan, other management reviews may be announced and held.

Preconditions. A management review should be conducted only when the objectives for the review have been established and all required inputs are available.

Management Preparation. Managers should ensure that the review is performed as planned and that appropriate time and resources are allocated in support of the review process. They should ensure that all participants are technically qualified and have received training and orientation in support of the review. They are responsible for ensuring that all reviews are carried out as planned and that any resulting action items are completed.

Planning the Review. The review leader is responsible for identifying the review team and the assignment of specific responsibilities. The leader is responsible for scheduling the meeting and for the distribution of all materials required in support of review preparation activities. Prereview comments should be collected by the review leader, classified, presented to the author prior to the review.

Overview of Review Procedures. A qualified person should present an overview of the review procedures for the review team if requested by the review leader.

Preparation. Each review team member should review the software product and any other review inputs prior to the review meeting. All anomalies detected during this examination should be documented and sent to the review leader.

Examination. The management review should consist of one or more meetings of the review team. The meetings should review the objectives of the review, evaluate the software product and product status, review all items identified prior to the review meeting, generate a list of action items, including associated risk. The meeting should be documented. This documentation should include any risk issues that are critical to the success of the project, provide a confirmation of the software requirements, list action items, and identify other issues that should be addressed.

Rework/Follow-up. The review leader is responsible for ensuring that all action items assigned in the meeting are closed.

Exit Criteria. The management review is complete when all required activities, outputs, and follow-up activities are finished.

Output. The output from the management review should identify:

    a. The project being reviewed
    b. The review team members
    c. Review objectives
    d. Software product reviewed
    e. Specific inputs to the review
    f. All action items
    g. A list of all anomalies

Audits

The information provided here is based upon the recommendations provided in IEEE Std 1028-1997, IEEE Standard for Software Reviews. Audits are an effective way to detect and identify problems in the software development process and to ensure that practices and procedures are being implemented as expected. Audits can also be performed by comparing what people are doing against established plans and procedures. The criteria for audits are presented here to help organizations define their review practices.

Introduction. A software audit provides an independent evaluation of conformance. Examples of software products subject to audit include the following:

    a. Backup and recovery plans
    b. Contingency plans
    c. Contracts
    d. Customer complaints
    e. Disaster plans
    f. Hardware performance plans
    g. Installation plans and procedures
    h. Maintenance plans
    i. Management review reports
    j. Operations and user manuals
    k. Procurement and contracting methods
    l. Reports and data
    m. Risk management plans
    n. Software configuration management plans

o. Software design descriptions

p. Source code

q. Unit development folders

r. Software project management plans

s. Software quality assurance plans

t. Software requirements specifications

u. Software safety plans

v. Software test documentation

w. Software user documentation

x. Software verification and validation plans

Responsibilities. The following roles should be established for an audit:

Lead auditor. The lead auditor is responsible for all administrative tasks, assembly of the audit team and their management, and for ensuring that the audit meets its objectives. The lead auditor should prepare a plan for the audit and summary activities with an audit report. The lead auditor should be free from bias and influence that could reduce any ability to make independent, objective evaluations.

Recorder. The recorder should document all anomalies, action items, decisions, and recommendations made by the audit team.

Auditor. The auditors should examine all products as described in the audit plan. They should record their observations and recommend corrective actions. All auditors should be free from bias and influences that could reduce their ability to make independent, objective evaluations.

Initiator. The initiator determines the need, focus, and purpose of the audit. The initiator determines the members of the audit team and reviews the audit plan and audit report.

Audited organization. The audited organization should provide a liaison to the auditors and is responsible for providing all information requested by the auditors. When the audit is completed, the audited organization should implement corrective actions and recommendations.

Input. Inputs to the audit should be listed in the audit plan and should include the purpose and scope of the audit, a list of products to be audited, and audit evaluation criteria.

Authorization. An initiator decides upon the need for an audit. A project milestone or a nonroutine event, such as the suspicion or discovery of a major nonconformance, may drive this decision. The initiator selects an auditee and provides the auditors with all supporting information. The lead auditor produces an audit plan and the auditors prepare for the audit.

Preconditions. An audit should only be conducted if the audit has proper authorization, the audit objectives have been established, and all audit items are available.

Procedures. The audit procedures are:

Management preparation. Managers are responsible for ensuring that the appropriate time and resources have been planned in support of the audit activities. Managers should also ensure that adequate training and orientation on audit procedures is provided.

Planning the audit. The audit plan should describe the purpose and scope of the audit, the audited organization, including location and management, and a list of software products to be audited. The plan should describe all evaluation criteria, auditor's responsibilities, examination activities, resource requirements and schedule, and required reporting. The initiator should approve the audit plan but the plan should be flexible enough to allow for changes based on information gathered during the audit, subject to approval by the initiator.

Opening meeting. An opening meeting between the audit team and audited organization should be scheduled to occur at the beginning of the examination activity of the audit. This meeting should cover the purpose and scope of the audit, all software products being audited, all audit procedures and outputs, audit requirements, and audit schedule.

Preparation. The initiator should notify the audited organization's management in writing before the audit is performed, except for unannounced audits. The purpose of notification is to ensure that the people and material to be examined in the audit are available.

Auditors should prepare for the audit by reviewing the audit plan and any information available about the audited organization and the products to be audited. The lead auditor should prepare for the team orientation and any necessary training. The lead auditor should also ensure that facilities are available for the audit interviews and that all materials are available.

Examination. The examination should consist of evidence collection and analysis with respect to the audit criteria as described in the audit plan. A closing meeting between the auditors and audited organization should be conducted, and then an audit report should be prepared.

Evidence Collection. The auditors should collect evidence of conformance and nonconformance by interviewing audited-organization staff, examining documents, and witnessing processes. The auditors should attempt all the examination activities defined in the audit plan and undertake additional activities if they consider them to be required to determine conformance. Auditors should document all observations of conformance or nonconformance. These observations should be categorized as major or minor. An observation should be classified as major if the nonconformity will likely have a significant effect on product quality, project cost, or project schedule. All observations should be discussed with the audited organization before the closing audit meeting.

Closing Meeting. The lead auditor should convene a closing meeting with the audited organization's management. The closing meeting should review the progress of the audit, all audit observations, and preliminary conclusions and recommendations. Agreements should be reached during the closing audit meeting and must be completed before the audit report is finalized.

Reporting. The lead auditor is responsible for the preparation of the audit report. The audit report should be prepared as quickly as possible following the audit. Any communication between auditors and the audited organization made between the closing meeting and the issue of the report should pass through the lead auditor. The lead auditor should send the audit report to the initiator who will distribute the audit report within the audited organization.

Follow-up. The initiator and audited organization should determine what corrective action may be required and the type of corrective action to be performed.

Exit Criteria. An audit is complete when the audit report has been delivered to the initiator and all audit recommendations have been performed.

Output. The output of the audit is the audit report. The audit report should describe the purpose and scope of the audit; the audited organization, including participants; identification of all products included in the audit; and recommendations. All criteria (e.g., standards, procedures) in support of the audit should be identified in the plan. A summary of all audit activities should be provided. All classified observations should be included, as well as a summary and interpretation of the findings. A schedule and list of all follow-up activities should be described.

PROCESS AND PRODUCT QUALITY ASSURANCE

Records collection, maintenance, and retention are directly addressed by IEEE Std 730, IEEE Standard for Software Quality Assurance Plans [3]. Emphasis is placed on these activities as described by the CMMI®-SW, SP 2.2—Establish and Maintain Records of the Quality Assurance Activities. These activities were implied by the SW-CMM® but not directly addressed in the manner that they are in the CMMI®.

IEEE Std 12207.0, Standard for Information Technology—Software Life Cycle Processes [39], describes 17 processes spanning the entire life cycle of a software product or service. Even if an organization's processes were defined using other sources, the standard is useful in characterizing the essential characteristics of these software processes and should be considered prior to the implementation of process improvement activities. Referencing this standard and reviewing what is required for each of these primary process areas can provide additional guidance in support of the activities associated with process and product quality assurance. It is important that the life cycle processes as described in IEEE 12207 be considered prior to the implementation of any process improvement activities associated with process and product quality assurance. The information provided within this section in support of the Process and Product Quality Assurance PA conforms with the information as described in IEEE Standards 12207.0 and 12207.1.

CMMI®-SW Goals

The Process and Product Quality Assurance process area provides for the definition of the activities associated with software project oversight. This PA supports all process areas by requiring the objective evaluation of conformance to stated project processes. Process and Product Quality Assurance ensures that the project staff and management have proper visibility into the processes and work products throughout the life cycle of a specified project.

SG1. Objectively Evaluate Processes and Work Products
Adherence of the performed process and associated work products and services to applicable process descriptions, standards, and procedures is objectively evaluated.

IEEE Std 730 can be used in conjunction with the IEEE Standard for Software Project Management Plans to develop processes and the supporting documentation required to meet the SG1 and SG2 goals for CMMI®-SW Process and Product Quality Assurance. These standards provide the details required to support software quality assurance (SQA) activities, including process development, plan documentation, and process and plan maintenance, evaluation, and modification.

SG2. Provide Objective Insight
Noncompliance issues are objectively tracked and communicated, and resolution is ensured.

Records collection, maintenance, and retention is directly addressed by IEEE Std 730. Emphasis is placed on these activities as described by the CMMI®-SW (Staged), SP 2.2— Establish and Maintain Records of the Quality Assurance Activities. These activities were implied by the SW-CMM® but not directly addressed in the manner that they are in the CMMI®.

CG2. Institutionalize a Managed Process
The process is institutionalized as a managed process.

The CMM® did not fully support the requirement to establish and maintain the software quality assurance plan (SQAP). The CMMI®-SW requires a plan for performing the process and product quality assurance process. This is different from just requiring the development of an SQAP. This CMMI®-SW requirement states that the process used to define SQA at the project level must be defined. There must be a project level SQAP, but also some type of organizational plan describing how this plan should be developed. IEEE Std 730 can be used to support both planning activities.

The following offers a suggested modification to the outline proposed by IEEE Std 730. In order to maximize quality, the application of improved processes reduce the number of defects that are introduced during software development. Table 5-6 describes the specific and generic goals and practices, typical work products, and commonly associated plans or artifacts in support of the PA.

Software Quality Assurance Plan

The information provided here is designed to facilitate the definition of processes and procedures relating to software quality assurance activities. This guidance was developed using IEEE Std 12207.0, Guide to Lifecycle Processes [39], and IEEE Std 730-2002, IEEE Standard for Software Quality Assurance Plans [3], which has been adapted to support CMMI requirements. The modification of the recommended SQAP table of contents to support the goals of the CMMI® more directly is shown in Table 5-7. Additional information is provided in Appendix C, Software Process Work Products, which presents a recommended minimum set of software reviews and an example SQA inspection log.

Table 5-6. Process and product quality assurance goals and practices [61]

Specific and generic goals/practices and typical work products	Plan or artifact
SG1. Objectively Evaluate Processes and Work Products	
SP 1.1. Objectively evaluate processes	
Evaluation reports	Software quality assurance plan
Noncompliance reports	Software quality assurance plan
Corrective actions	Software quality assurance plan
SP 1.2. Objectively evaluate work products and services	
Evaluation reports	Software quality assurance plan
Noncompliance reports	Software quality assurance plan
Corrective actions	Software quality assurance plan
SG2. Provide Objective Insight	
SP 2.1. Communicate and ensure resolution of noncompliance issues	
Corrective action reports	Software quality assurance plan
Evaluation reports	Software quality assurance plan
Quality trends	Software quality assurance plan
SP 2.1. Establish records	
Evaluation logs	Software quality assurance plan
Quality assurance reports	Software quality assurance plan
Status reports of corrective actions	Software quality assurance plan
Reports of quality trends	Software quality assurance plan
GG2. Institutionalize a Managed Process	
GP 2.1. Establish an organizational policy	Organizational policy that for planning and performing the product and process quality assurance process
GP 2.2. Plan the process	Software quality assurance plan
GP 2.3. Provide resources	
Identification of resources used in support of quality assurance activities	Software quality assurance plan
GP 2.4. Assign responsibility	Software project management plan, software quality assurance plan
GP 2.5. Train people	Software project management plan or training plan
GP 2.6. Manage configurations	
Description of how work products and all associated artifacts in support of process and product quality assurance activities are placed under configuration management	Configuration management plan
GP 2.7. Identify relevant stakeholders	
Describe measurement analysis feedback mechanisms	Software quality assurance plan
GP 2.8. Monitor and control the process	Software quality assurance plan

Table 5-6. *Continued*

Specific and generic goals/practices and typical work products	Plan or artifact
GG2. Institutionalize a Managed Process (*cont.*)	
GP 2.9. Objectively evaluate adherence	
Reviews are identified in SQA plan	Software quality assurance plan
GP 2.10. Review status with higher-level management	Software project management plan and software quality assurance plan
GG3. Institutionalize a Defined Process	
GP 3.1. Establish a defined process	
Establish and maintain a description of the process and product quality assurance process	Software quality assurance plan
GP 3.2. Collect improvement information	Requirements management plan, measurement and metrics plan, or software project management plan

Table 5-7. Software quality assurance plan document outline

Title Page
Revision Page
Table of Contents
1. Introduction
   1.1 Purpose
   1.2 Scope
   1.3 Definitions, Acronyms, and Abbreviations
   1.4 References
2. SQA Management
   2.1 Organization
   2.2 Tasks
   2.3 Responsibilities
3. SQA Documentation
   3.1 Development, Verification and Validation, Use, and Maintenance
   3.2 Control
4. Standards and Practices
   4.1 Coding/Design Language Standards
   4.2 Documentation Standards
5. Reviews and Audits
6. Configuration Management
7. Testing
8. Software Metrics
9. Problem Tracking
10. Records Collection, Maintenance, and Retention
11. Training
12. Risk Management

Software Quality Assurance Plan Document Guidance

The following provides section-by-section guidance in support of the creation of a SQAP. The SQAP should be considered to be a living document and should change to reflect any process improvement activity. This guidance should be used to help define a software quality process and should reflect the actual processes and procedures of the implementing organization. All information is provided for illustrative purposes only. Additional information is provided in the document template, *Software Quality Assurance Plan.doc,* which is located on the CD-ROM accompanying this book.

Introduction. This section should provide the reader with a basic description of the project, including associated contract information and any standards used in the creation of the document. The following is provided as an example:

> The [Project Name] ([Proj Abbrev]) software products are developed under contract # [contract number], Subcontract # [sub number], controlled by [customer office designator] at [customer location]. This Software Quality Assurance Plan (SQAP) has been developed to ensure that [project abbreviation] software products conform to established technical and contractual requirements. These may include, but are not limited to [company name] standards for Software Quality Assurance Plans.

As described by IEEE Std 610.12, Software Quality Assurance (SQA) is "a planned and systematic pattern of all actions necessary to provide adequate confidence that an item or product conforms to established technical requirements" [2]. This definition could be interpreted somewhat narrowly. The CMMI supports the notion that quality assurance is performed on both processes and products, and the 610.12 definition mentions only an item or product. This definition should be expanded to read "a planned and systematic pattern of all actions necessary to provide adequate confidence that an item, product, or process conforms to established technical requirements."

Purpose. This section should describe the purpose of the SQAP. It shall list the name(s) of the software items covered by the SQAP and the intended use of the software. An example is provided:

> The purpose of this SQAP is to establish, document, and maintain a consistent approach for controlling the quality of all software products developed and maintained by the [project abbreviation] Group. This plan defines the standardized set of techniques used to evaluate and report upon the process by which software products and documentation are developed and maintained.
>
> The SQA process described applies to all development, deliverables, and documentation maintained for the [project abbreviation] project. The SQA process and how it is implemented in the production, or maintenance life cycle, of each software product is shown in Figure xx.

Scope. This section should describe the specific products and the portion of the software life cycle covered by the SQAP for each software item specified. The following is an example of scope:

> The specific products developed and maintained by the [PROJECT ABBREVIATION] development team may be found in the [project abbreviation] Software Configuration Management Plan.

In general, the software development products covered by this SQA Plan include, but are not limited to, the following:

- Source files
- Libraries
- Executable files
- Data files
- Makefiles
- Link files
- Documentation

The software deliverables covered by this SQA Plan are:

- [Project name] System Software
- Version Description Documents (VDDs)
- Users Manuals
- Programmer's Manuals
- Software Design Documents
- Software Requirements Specifications

[project abbreviation] software development tasks cover the full spectrum of software life cycle stages from requirements to system testing. As a result, Software Quality Assurance (SQA) activities must be adapted as needed for each specific software task.

Because of the cumulative and diverse nature of [PROJECT ABBREVIATION] software, this life cycle often includes the use of programs or subprograms that are already in use. These programs are included in the Preliminary Design, test documentation, and testing. They are included in the Detailed Design strictly in the design or modification of their interfaces with the remainder of the newly developed code.

In many of the [PROJECT ABBREVIATION] software products, changes are required to existing code, either as corrections or upgrades. The life cycle of software change is the same as the life cycle of new software, although certain steps may be significantly shorter. For example, system testing may omit cases that do not exercise the altered code.

The decision as to whether an individual task involves a new software development or a change to an existing program will be the responsibility of the Project Lead, Program Manager, Customer, associated CCB, or a combination thereof.

The life cycle for a specific project would be described in the associated Software Project Management Plan (SPMP). An example of a project life cycle and information in support of SQA activities is provided in Appendix C, Example Life Cycle, of this text.

SQA Management. This section should describe the project's quality assurance organization, its tasks, and its roles and responsibilities, and will be used to effectively implement process and product quality assurance. A reference to any associated software project management plan should be provided here. The following is an example paragraph:

The [PROJECT ABBREVIATION] Program Manager, each [PROJECT ABBREVIATION] Project Team Lead, and the [PROJECT ABBREVIATION] Software Quality Assurance Manager (SQAM) will work together with the [Company Name] Software Engineering Process Group to develop, maintain, and implement effective software quality assurance management.

Organization. This section should depict the organizational structure that influences and controls the quality of the software. This should include a description of each major element of the organization together with the roles and delegated responsibilities. The amount of organizational freedom and objectivity to evaluate and monitor the quality of the software, and to verify problem resolutions, should be clearly described and documented. In addition, the organization responsible for preparing and maintaining the SQAP should be identified. A typical organizational breakdown follows:

Program Manager. The [PROJECT ABBREVIATION] Program Manager (PM) is ultimately responsible for ensuring the SQA process is developed, implemented, and maintained. In this capacity, the PM will work with the [COMPANY NAME] SQAM and the [PROJECT ABBREVIATION] SQAM in establishing standards and implementing the corporate SQAP as it applies to the [PROJECT ABBREVIATION] program.

The [PROJECT ABBREVIATION] PM is ultimately responsible for ensuring that the Project Lead(s) and their associated project team(s) are trained in SQA policies and procedures that apply to their function on each team. This will consist of initial training for new members and refresher training, as required, due to process or personnel changes.

Project Lead. The [PROJECT ABBREVIATION] Project Lead (PL) shall provide direct oversight of SQA activities associated with project software development, modification, or maintenance tasks assigned within project functional areas.

The Project Lead(s) are responsible for ensuring that project software quality assurance processes are followed. Similarly, the PL will ensure that all members of the project team regularly implement the SQA policies and procedures that apply to their specific function on the team.

The PL is responsible for implementing the actions of this SQA process for all [PROJECT ABBREVIATION] software products, including end-user products, module products, and data. The PL will be responsible for controlling the product, ensuring that only approved changes are implemented. Each PL will maintain this process, with all changes to the process approved by the PM and associated management.

Tasks. This section should describe the portion of the software life cycle covered by the SQAP. All tasks to be performed, the entry and exit criteria for each task, and relationships between these tasks and the planned major checkpoints should also be described. The sequence and relationships of tasks, and their relationship to the project management plan master schedule, should also be indicated. The following provides a content example:

The [PROJECT ABBREVIATION] PM shall ensure that a planned and systematic process exists for all actions necessary to provide software products that conform to established technical and contractual requirements.

The following SQA checkpoints will be placed in the software development life cycle:

  a. At the end of the Software Requirements and Software System Test Planning activities, there shall be a Requirements Review in accordance with Section xxx of this document.
  b. At the end of the Preliminary Design activity, there shall be a Preliminary Design Review in accordance with Section xxx of this document.
  c. At the end of Detailed Design, there shall be a Critical Design Review in accordance with Section xxx of this document.
  d. At the end of Unit Testing, there shall be a walkthrough in accordance with Section xxx of this document.

e. At the end of Integration Testing, there shall be a Test Review in accordance with Section xxx of this document.

f. At the end of System Testing, there shall be a Test Review and an audit in accordance with Section xxx of this document.

g. At Release, there shall be an SQA review using the checklists found in Appendix xx.

The following SQA checkpoints will be placed in the software maintenance life cycle:

a. At the end of the Analysis activity there shall be a Requirements Review in accordance with Section xxx of this document.

b. At the end of Design, there shall be a Preliminary Design Review in accordance with Section xxx of this document.

c. At the end of Implementation, there will be an walkthrough in accordance with procedures identified in the associated System Test Plan and in Section xxx of this document.

d. At Release, there shall be an SQA review using the release checklists found in Appendix xx.

e. Acceptance Testing is dependant upon criteria determined by the customer. This criterion is based upon the documentation released with the software (i.e., the SRS, VDD, users documentation, etc.).

Each PL, or a designee, shall evaluate all software products to be delivered to the customer during the course of any project. This evaluation must ensure that the products are properly annotated, are adequately tested, are of acceptable quality, and are updated to reflect changes specified by the customer. Additionally, the PL is responsible for ensuring that both the content and format of the software documentation meet customer requirements.

Responsibilities. This section should identify the specific organizational element that is responsible for performing each task, as in the following:

Project Lead Responsibilities. The [PROJECT ABBREVIATION] PM has overall responsibility for [PROJECT ABBREVIATION] program SQA responsibilities.

The [PROJECT ABBREVIATION] SQAM has the overall responsibility for monitoring compliance with the project procedures, noting any deviations and findings reporting.

The [PROJECT ABBREVIATION] PLs have the responsibility for ensuring that individual team members are trained in Project SQA procedures and practices. They may delegate the actual training activities to the [PROJECT ABBREVIATION] SQAM. Project Leads are also responsible or reporting changes in Project procedures to the [PROJECT ABBREVIATION] SQAM.

Each individual [PROJECT ABBREVIATION] employee shall be responsible for learning and following the SQA procedures and for suggesting changes to their PL when necessary. Each PL shall be responsible for the implementation and accuracy of all SQA documentation and life cycle tracking.

Configuration Control Board (CCB). Each baseline software product developed by [PROJECT ABBREVIATION] has an associated Configuration Control Board (CCB).

The associated CCB has the authority to approve, disapprove, or revise all Software Quality Assurance reporting procedures.

An example of the delineation of SQA life cycle responsibility is provided in Appendix C, Example Life Cycle, of this text.

SQA Documentation. This section should identify the documentation governing the development, verification and validation, use, and maintenance of the software. List all documents that are to be reviewed or audited for adequacy, identifying the reviews or audits to be conducted. Refer to IEEE Std 730-2002 for detailed information supporting SQA documentation requirements.

Development, Verification and Validation, Use, and Maintenance. This section should address how project documentation is developed, verified, validated, and maintained.

Control. This section should describe the responsibilities associated with software project documentation. The following provides an example of the type of information to support this requirement:

> The [PROJECT ABBREVIATION] ([PROJECT ABBREVIATION]) Program Manager (PM), Project Lead(s), and team members shall share responsibility for SQA documentation. Document reviews and integration of any required documentation changes will occur as specified in Section 2.3.1 of this document. Refer to *[PROJECT ABBREVIATION] Software Configuration Management Process* for additional information regarding specific product (i.e., software, documentation . . .) maintenance.
>
> Changes to requirements, software baselines, and completed maintenance items will be reported in the deliverable's Version Description Document. Status reports will be provided to [Customer Name] no less than biweekly.

Standards and Practices. This section should identify the standards, practices, conventions, quality requirements, and metrics to be applied. Product and process measures should be included in the metrics used and may be identified in a separate metrics plan. This section should also describe how conformance with these items will be monitored and assured. The following provides an example of typical information:

> The following standards shall be applied to all programs and incorporated into existing programs whenever and wherever possible. Retrofitting of existing programs to meet new standards will be avoided when this retrofitting impacts production.
>
> Coding/Design Language Standards. All software developed and maintained by [PROJECT ABBREVIATION] project teams shall adhere to *[PROJECT ABBREVIATION] Programming Standards, [Additional Coding Reference].* Deviations from the Standards shall be noted on the SCR form or recorded as a CER.
>
> *Documentation Standards.* All [PROJECT ABBREVIATION] documentation shall be adapted from the [Additional Documentation Reference] and IEEE documentation guidelines.

Reviews and Audits. This section should describe all software reviews to be conducted. These might include managerial reviews, acquirer–supplier reviews, technical reviews, inspections, walk-throughs, and audits. The schedule for software reviews as they relate to the software project's schedule should also be listed, or reference made to a Software Project Management Plan containing the information. This section should describe how the software reviews would be accomplished.

Configuration Management. This section should identify associated configuration management activities necessary to meet the needs of the SQAP. An example follows:

For information regarding configuration control and status accounting procedures, refer to the *[PROJECT ABBREVIATION] Configuration Management Plan*. Configuration Management Maintenance is the responsibility of each [PROJECT ABBREVIATION] PL and each project team Configuration Manager.

The Project SQA Manager shall audit the baseline versions of the software product and accompanying documentation prior to software release for validation. (Please refer to Appendix xx, Software Checklists). The [PROJECT ABBREVIATION] PM will verify that a baseline is established and preserved and that documented configuration management policies and [COMPANY NAME] corporate policies are implemented.

Testing. This section should identify all the tests not included in the software verification and validation plan for the software covered by the SQAP, and should state the methods to be used. If a separate test plan exists, it should be referenced. An example follows:

> Testing of software developed, modified and/or maintained under the [PROJECT ABBREVIATION] contract must be tailored to meet the requirements of the specific project Work Request. Typical testing efforts include unit, integration, and system testing. The [PROJECT ABBREVIATION] Program Manager will ensure that all project software testing is planned and executed properly. Please refer to the *[PROJECT ABBREVIATION] Configuration Management Process* and program test plans, listed in Section xxx.

The system test program, also referred to as DT&E testing, should be broad enough to ensure that all the software requirements are met. The Software Test Plan is to be written at the same time that the software requirements are being finalized. It may be useful to draft a test plan while the requirements are being defined. As each requirement is defined, a way to determine that it is satisfied should be delineated. This method is then translated into a test case for the system testing and added to the Software Test Plan.

System Testing. System testing should consist of a list of test cases (descriptions of the data to be entered) and a description of the resulting program output.

Integration Testing. Integration testing should check to ensure that the modules work together as specified in the SDD. Integration testing should be included as part of the Software Test Plan, ensuring that all the modules to be tested are executed and exercised thoroughly.

Unit Testing. Unit testing is the detailed testing of a single module to ensure that it works as it should. Unit testing should be identified in the Software Test Plan to ensure that every critical function in the module is executed at least once. Critical functionality should be identified in the Software Test Plan.

Software Metrics. This section should describe the collection, maintenance, and retention of metrics associated with the software quality assurance process. A separate software measurement and metrics plan is often used, and if so, a reference to this plan should be provided. The following is an example of the information supporting this section of the document:

> The [PROJECT ABBREVIATION] ([PROJECT ABBREVIATION]) SQA Manager shall ensure that [PROJECT ABBREVIATION] Software projects follow the guidelines established by the *[PROJECT ABBREVIATION] Software Measurement and Metrics Plan*.
>
> The SQA Manager shall ensure the project metrics are appropriate and will provide recommendations for improvements on the project's measurement process, as required.

The following software metrics will be recorded and used in support of the [PROJECT ABBREVIATION] SQA effort:

a. A history of the bugs and enhancements, a list of functions which are changed to fix the bugs or implement the enhancements, and the time and effort required to fix the bugs or implement the enhancements; the [PROJECT ABBREVIATION] Software Change Request Tracking System will provide such information (CER/SCR System).

b. Total numbers of errors, opened versus closed; errors opened and closed since the last report. Errors being defined as:

   1. Requirements errors—Baseline Change Request (BCR)
   2. Coding defects—Test Problem Reports (TPRs)
   3. Documentation defects—Interface Change Notice (DCR)
   4. Bad Fixes—Internal Test Report (ITR)

c. Total implemented requirements vs planned requirements.]

Problem Tracking. This section should describe the practices and procedures to be followed for reporting, tracking, and resolving problems or issues identifieded in both software items and the software development and maintenance process. It should also describe organizational responsibilities concerned with their implementation. The following provides some example text:

Please refer to *[PROJECT ABBREVIATION] Configuration Management Process.* This document describes the practices and procedures followed for reporting, tracking, and resolving problems identified in software items and the associated development, maintenance, and organizational responsibilities.

Deficiencies in the SQA Plan shall be corrected using the following procedure:

1. A deficiency is noted, either by a [COMPANY NAME] employee finding the procedure difficult to follow or by the SQA manager noting that the procedure has not been followed.

2. The PM, or a designee, notes the deficiency and develops a way to correct it. In this, he shall have help from the employee who had trouble following the procedure or who did not follow the procedure.

3. A new version of the SQA Plan is written, printed and disseminated to [PROJECT ABBREVIATION] Program personnel and [COMPANY NAME] Corporate SQA. The new version shall incorporate the correction.

Records Collection, Maintenance, and Retention. This section should identify all SQA documentation and describe how the documentation is maintained while designating the retention period. The following is an example:

The Software Developmental Library (SDL) for [PROJECT ABBREVIATION] program software may be found in Appendix A. Documents pertaining to developmental configuration, which are also maintained, may be found in Section 3.1. Access to all documents is controlled by each project team's Project Lead, or designee.

This documentation will be maintained in [Maintenance location and address]. All documentation will be maintained on in a shared [PROJECT ABBREVIATION] subdirectory [PROJECT ABBREVIATION]/[path]/shared/<document name>. Weekly backups of this system occur and are stored off-site.

All documentation will be maintained throughout the length of the [PROJECT ABBRE-VIATION] contract.

Training. This section should identify the training activities necessary to meet the needs of the SQAP. The following is an example:

It is the responsibility of the [PROJECT ABBREVIATION] Program Manager (PM) to allow for the training necessary to accomplish SQA goals and implementation. Similarly, the PL will ensure that all members of the project team are trained in the SQA policies and procedures that apply to their function on each team. This will consist of initial training for new members and refresher training, as required, due to process or personnel changes.

Risk Management. This section should specify all methods and procedures employed to identify, assess, monitor, and control areas of risk arising during the portion of the software life cycle covered by the SQAP. Reference to the software project management plan, or a separate risk management plan, would suffice here, as shown by the following text:

Program risks will be identified either by individual team members or CCB members. The risk tracking form (refer to [PROJECT ABBREVIATION] Software Development Plan) will be used to document and track all risks identified. All risks will be submitted to the [PROJECT ABBREVIATION] CCB at weekly staff meetings to determine the severity and probability. Risks determined to be significant will be assigned an owner. The owner will determine the plan for dealing with the risk and propose a plan to the [PROJECT ABBREVIATION] CCB. The owner will track the risk to closure. Risks that are not deemed significant will be canceled.

The guidelines used by the [PROJECT ABBREVIATION] CCB to define risk severity and probability are listed in the [PROJECT ABBREVIATION] Software Development Plan.

Additional information is provided in the *Software Quality Assurance Plan.doc* document template on the CD-ROM accompanying this book. Appendix C, Software Process Work Products, of this book also provides several examples of supporting process artifacts.

CONFIGURATION MANAGEMENT

Configuration management comprises the processes in support of the definition, control, review, and reporting of the work products associated with a software project. This PA ensures that the integrity of all work products, and all items used to create any work product, is established and effectively maintained. Configuration management (CM) activities ensure that changes to work products are captured and managed. This PA applies to all other process areas by requiring that all the processes, and process artifacts, used to support a software effort be placed under configuration management control. CM activities include the management of requirements baselines and supporting project plan revision management and source code control.

Standards provide written expectations for the processes to be used and the deliverable to be produced by software developers. IEEE Std 828, IEEE Standard for Software Configuration Management Plans [4], and IEEE Std 1042,* ANSI/IEEE Guide to Software Configuration Management [1], may be used to highlight the specific areas of configura-

*IEEE Std 1042 has expired, but is still available from IEEE Press upon request.

tion management that should be considered for process development and documentation. IEEE Std 828 directly supports most of the Level 2 CMMI®-SW CM PA requirements and can be used with minor modification. It is important to note that this plan should be used in conjunction with the software quality assurance plan, which would address items like the types of reviews to be performed on the CM plan, and the software project management plan, which would address other supporting plans.

IEEE Std 12207.0, Standard for Information Technology—Software Life Cycle Processes [39], describes 17 processes spanning the entire life cycle of a software product or service. Even if an organization's processes were defined using other sources, this standard is useful in characterizing the essential characteristics of these software processes and should be considered prior to the implementation of process improvement activities. Referencing this standard and reviewing what is required for each of these primary process areas can provide additional guidance in support of the activities associated with configuration management. It is important that IEEE 12207 be considered prior to the implementation of any process improvement activities associated with configuration management. The information provided within this section in support of the Configuration Management PA conforms to with the information as described in IEEE Stds 12207.0 and 12207.1.

CMMI®-SW Goals

The CMMI®-SW [61] emphasizes the need to place acquired products under configuration management by both the supplier and the project. If required, this can be supported by IEEE Std 1062-1998, Recommended Practice for Software Acquisition. Configuration management of work products may be performed at several levels of granularity. Provisions for conducting configuration management should be established in supplier agreements.

SG 1. Establish Baselines
Baselines of identified work products are established.

IEEE Std 828-1998, IEEE Standard for Software Configuration Management Plans, requires that a configuration management system be established and that the identification of all configuration items and all release baselines must be established. The CMMI®-SW identifies a baseline as the set of requirements, design, source code files and the associated executable code, build files, and user documentation that have been assigned a unique identifier. Release of a baseline requires retrieval of source code files from the configuration management system.* IEEE Std 828 effectively supports the CMMI specific goal for establishing baselines.

SG 2. Track and Control Changes
Changes to the work products under configuration management are tracked and controlled.

IEEE Std 828 requires that once baselines are established, that they be effectively maintained. Baselines are not static. For example, changes to product baselines occur through requirements implementation. Phase 1 of a project may be complete, so the product is

---

*A baseline that is delivered to a customer is typically called a "release" whereas a baseline for an internal use is typically called a "build."

baselined. Then phase 2 begins, requiring the development of a new product baseline. Baselines change may also be due to customer requests or changes in requirements. IEEE Std 828 supports effective baseline management; all changes to items that collectively comprise the product baseline must be tracked.

SG 3. Establish Integrity
Integrity of baselines is established and maintained.

IEEE Std 828 requires that each product baseline be identified for recovery, if necessary. It requires the description of all items included in a product baseline. This description must include a complete revision history describing any changes and associated change rationale. Configuration audits are also required in support of baseline maintenance once the baseline is established.

GG 2. Institutionalize a Managed Process
The process is institutionalized as a managed process.

IEEE Std 828 requires that the individuals responsible for configuration management activities be assigned specific responsibilities and that they be granted authority to accomplish the goals of this PA. The importance of adequate training in support of configuration management activities is also emphasized. It is important to note that this plan should be used in conjunction with quality assurance activities, which would address items like the types of reviews to be performed on the configuration management plan, and the project management plan that would address other supporting plans. Table 5-8 describes the specific and generic goals and practices, typical work products, and commonly associated plans or artifacts in support of the PA.

Software Configuration Management Plan

The CMMI®-SW fully supports the requirement to establish and maintain a plan for performing all configuration management process activities. The CMMI®-SW also requires the documentation of project-level CM activities, and also requires the description of organizational CM activities. IEEE Std 828-1998, IEEE Standard for Software Configuration Management Plans (SCMP), can be used to help support this requirement. Appendix C also provides examples of supporting CM work products, which include a Configuration Control Board (CCB) Letter of Authorization, CCB Charter, and Software Change Request Procedures. The modification of the recommended CM table of contents to support the goals of the CMMI® more directly is shown in Table 5-9.

Software Configuration Management Plan Document Guidance

The following provides section-by-section guidance in support of the creation of a software configuration management plan (SCMP). The SCMP should be considered to be a living document and should change to reflect any process improvement activity. This guidance should be used to help define a software configuration management process and should reflect the actual processes and procedures of the implementing organization. Additional information is provided in the document template, *Software Configuration Management Plan.doc,* which is located on the CD-ROM accompanying this book.

Table 5-8. Configuration management goals and practices [61]

Specific and generic goals/practices and typical work products	Plan or artifact
SG1. Establish Baselines	
SP 1.1. Identify configuration items	
Identified configuration items	Configuration management plan
SP 1.2. Establish a configuration management system	
Configuration management system with control work products	Configuration management plan
Configuration management system access control procedures	Configuration management plan
Change request database	Configuration management plan
SP 1.3. Create or release baselines	
Baselines	Configuration management plan
Description of baselines	Configuration management plan
CCB procedures; SCM plan	Configuration management plan
SG2. Track and Control Changes	
SP 2.1. Track change requests	
Change requests	Configuration management plan
SP 2.2. Control configuration items	
Revision history of configuration items	Configuration management plan
Archives of the baselines	Configuration management plan
SG3. Establish Integrity	
SP 3.1. Establish configuration management records	
Revision history of configuration items	Configuration management plan
Change log	Configuration management plan
Copy of the change requests	Configuration management plan
Status of configuration items	Configuration management plan
Differences between baselines	Configuration management plan
SP 3.2. Perform configuration audits	
Configuration audit results	Configuration management plan
Action items	Configuration management plan
GG2. Institutionalize a Managed Process	
GP 2.1. Establish an organizational policy	Organizational policy for planning and performing the configuration management process
GP 2.2. Plan the process	Configuration management plan
GP 2.3. Provide resources	
Identification of resources used in support of configuration management activities	Configuration management plan
GP 2.4. Assign responsibility	Software project management plan or configuration management plan

Table 5-8. *Continued*

Specific and generic goals/practices and typical work products	Plan or artifact
GG2. Institutionalize a Managed Process (*cont.*)	
GP 2.5. *Train people*	Software project management plan or training plan
GP 2.6. *Manage configurations* Description of how work products and all associated artifacts in support of software configuration management activities are placed under configuration management	Configuration management plan
GP 2.7. *Identify relevant stakeholders* Describe activities and feedback mechanisms	Configuration management plan, organizational policy or software project management plan
GP 2.8. *Monitor and control the process*	Configuration management plan
GP 2.9. *Objectively evaluate adherence*	Software quality assurance plan
GP 2.10. *Review status with higher-level management*	Software project management plan
GG3. Institutionalize a Defined Process	
GP 3.1. *Establish a defined process* Establish and maintain a description of the configuration management process	Configuration management plan
GP 3.2. *Collect improvement information*	Requirements management plan, software measurement and metrics plan, or software project management plan

Introduction. The introduction information provides a simplified overview of the software configuration management (SCM) activities so that those approving, those performing, and those interacting with the SCM can obtain a clear understanding of the plan. The introduction should include four topics: the purpose of the plan, the scope, the definition of key terms, and references.

Purpose. The purpose should address why the plan exists and the intended audience.

Scope. The scope should address SCM applicability, limitations, and assumptions on which the plan is based. The scope should provide an overview description of the software development project, identification of the software CI(s) to which SCM will be applied, identification of other software to be included as part of the plan (e.g., support or test software), and relationship of SCM to the hardware or system configuration management activities for the project. The scope should address the degree of formality, depth of control, and portion of the software life cycle for applying SCM on this project, including any limitations, such as time constraints, that apply to the plan. Any assumptions that might have an impact on the cost, schedule, or ability to perform defined

Table 5-9. Software configuration management plan document outline

Title Page
Revision Page
Table of Contents
1.0   Introduction
    1.1   Purpose
    1.2   Scope
    1.3   Definitions/Acronyms
    1.4   References
2.0   Software Configuration Management
    2.1   SCM Organization
    2.2   SCM Responsibilities
    2.3   Relationship of CM to the software process life cycle
        2.3.1   Interfaces to other organizations on the project
        2.3.2   Other project organizations CM responsibilities
    2.4   SCM Resources
3.0   Software Configuration Management Activities
    3.1   Configuration Identification
        3.1.1   Specification identification
        3.1.2   Change control form identification
        3.1.3   Project baselines
        3.1.4   Library
    3.2   Configuration Control
        3.2.1   Procedures for changing baselines
        3.2.2   Procedures for processing change requests and approvals
        3.2.3   Organizations assigned responsibilities for change control
        3.2.4   Change control boards (CCBs)
        3.2.5   Interfaces
        3.2.6   Level of control
        3.2.7   Document revisions
        3.2.8   Automated tools used to perform change control
    3.3   Configuration Status Accounting
        3.3.1   Storage, handling and release of project media
        3.3.2   Information and control
        3.3.3   Reporting
        3.3.4   Release process
        3.3.5   Document status accounting
        3.3.6   Change management status accounting
    3.4   Configuration Audits and Reviews
4.0   Configuration Management Milestones
5.0   Training
6.0   Subcontractor Vendor Support

SCM activities (e.g., assumptions of the degree of customer participation in SCM activities or the availability of automated aids) must also be addressed. The following is an example of scope:

> This document defines CM activities for all software and data produced during the development of the [Project Name] ([Project Abbreviation]) software. This document applies to all module products, end-user products, and data developed and maintained for the [Project Ab-

breviation] Program. CM activities as defined herein will be applied to all future [Project Abbreviation] projects.

This document conforms to [Company Name]'s Software Configuration Management Policy [Policy #] and IEEE standards for software configuration management, and will change as needed to maintain conformance.

Software Configuration Management (SCM). Appropriately documented SCM information describes the allocation of responsibilities and authorities for SCM activities to organizations and individuals within the project structure. SCM management information should include three topics: the project organization(s) within which SCM is to apply, the SCM responsibilities of these organizations, and references to the SCM policies and directives that apply to this project.

SCM Organization. The SCM organizational context must be described. The plan should identify the all participants or those responsible for any SCM activity on the project. All functional roles should be described, as well as any relationships to external organizations. Organization charts, supplemented by statements of function and relationships, can be an effective way of presenting this information. An example is shown in Figure 5.1, which provides an overview of [Project Abbreviation] SCM organization.

SCM Responsibilities. All those responsible for SCM implementation and performance should be described in this section. A matrix describing SCM functions, activities, and tasks can be useful for documenting the SCM responsibilities. For any review board or special organization established for performing SCM activities on this project, the plan should describe its

    a. Purpose and objectives
    b. Membership and affiliations
    c. Period of effectivity
    d. Scope of authority
    e. Operational procedures

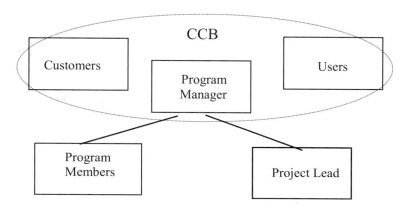

Figure 5.1. [Project Abbreviation] SCM organization.

The following provides an example of typical responsibilities:

The Program Manager (PM) is responsible for ensuring that the SCM process is developed, maintained, and implemented. The PM is responsible for ensuring that project leads and team members are adequately trained in SCM policy and procedure. The [Project Abbreviation] Program Manager and Project Leads work together to develop, maintain, and implement effective software configuration management. Please refer to Figure 3.1 and 3.1.

The [Project Abbreviation] Program Manager (PM) has the authority to ensure development of an appropriate configuration management process. Since CCB responsibility exists with the customer (See Figure 3.1), team member(s) of the [Project Abbreviation] Project will be identified as members of this CCB and will represent our interests there.

The Configuration Control Board (CCB) is responsible for the review and approval of all Software Change Requests (SCRs), all baseline items, and all changes to baseline items. The CCB considers the cost and impact of all proposed changes, and considers the impact to all interfacing items as part of their approval action. The CCB for [Project Abbreviation] consists of the PM, the PL, and two or more members from the government Program Management team.

The associated Configuration Control Board (CCB) has the authority to approve a baseline version of this process, as well as any changes to the process as recommended by any associated CCB representative. The same CCB has the authority to define the baseline for each software module, and end-user products, and to approve start of work and acceptance of completed changes to the baseline products. This CCB also has the authority to approve reversion to a previous version of the product if warranted.

The [Project Abbreviation] Program SCM is responsible for implementing the actions of this process and for the implementation of [Company Name] process improvement recommendations.

The Software Configuration Manager (SCM) ensures that the process is implemented once approved and that proposed changes to the process are reviewed by [Project Abbreviation] team members and approved by the CCB as required.

Project Leads, or a designee, are responsible for adhering to the [Project Abbreviation] CM process and for ensuring that only approved changes are entered into the module product baselines. Project Leads, or a designee, are also responsible for identifying the module version to be included in the end-user products. The Project Lead, or a designee, is responsible for controlling the software product(s) development and/or maintenance baseline to ensure that only approved changes are entered into the baseline(s) system. The Project Lead maintains the software product with all changes to be approved by the associated CCB.

The Project Lead (PL) develops and proposes a baseline version of the software configuration management product, ensuring that the SCM process is followed, once approved, and proposes process changes to the [Project Abbreviation] SCM and CCB, as required.

Each Project Lead (PL) is responsible for ensuring that project configuration management processes are followed. The PL also ensures all members of the project team are trained in SCM policies and procedures that apply to their function on the team. This consists of initial training for new members and refresher training, as required, due to process or personnel changes.

Relationship of CM to the Software Process Life Cycle. This section should describe CM as it relates to other elements of the software process life cycle. A diagram showing the relationship between CM elements and the overall process is often used here.

SCM Resources. This section should describe the minimum recommended resources (e.g., time, tools, materials) dedicated to software configuration management activities.

Configuration Identification. All technical and managerial SCM activities should be described, as well as all functions and tasks required to manage the configuration of the software. General project activities that have SCM implications should also be described from the SCM perspective. SCM activities are traditionally grouped into four functions: configuration identification, configuration control, status accounting, and configuration audits and reviews. An example of the type of information provided in this section follows:

> All [Project Abbreviation] Modules and baselines are to be approved by the CCB. Once a module has been baselined, all changes made will be identified in their respective Version Description Documents (VDDs).
>
> The [Project Abbreviation] is currently in development; product baseline to be approved upon delivery.
>
> Product baselines are to be maintained by the [Project Abbreviation] SCM. The baseline will consist of the following:
>
> - Source files
> - Libraries
> - Executable files
> - Data files
> - Makefiles
> - Link files
> - VDDs
> - Users Manuals

Specification Identification. SCM activities should identify, name, and describe all code, specifications, design, and data elements to be controlled for the project. Controlled items may include executable code, source code, user documentation, program listings, databases, test cases, test plans, specifications, management plans, and elements of the support environment (such as compilers, operating systems, programming tools, and test beds).

Change Control Form Identification. The plan should specify the procedures for requesting a change. As a minimum, the information recorded for a proposed change should contain the following:

a. The name(s) and version(s) of the CIs where the problem appears
b. Originator's name and organization
c. Date of request
d. Indication of urgency
e. The need for the change
f. Description of the requested change

Additional information, such as priority or classification, may be included to assist in its analysis and evaluation. Other information, such as change request number, status, and disposition, may be recorded for change tracking. An example of the type of information provided in this section follows:

> Software Change Request (SCR) procedures for each development team are defined in Appendix A. These procedures are used to add to, change, or remove items from the baselines.

The identification and tracking of change requests is accomplished through Change Enhancement Requests (CER).

Project Baselines. This section should describe the configuration control activities required to implement changes to baselined CIs. The plan should define the following sequence of specifc steps:

    a. Identification and documentation of the need for a change

    b. Analysis and evaluation of a change request

    c. Approval or disapproval of a request

    d. Verification, implementation, and release of a change

The plan should identify the records to be used for tracking and documenting this sequence of steps for each change.

Library. This section should identify the software configuration management repository, if one exists. The following provides an example of the type of typical information:

> [Project Abbreviation] is composed of end-user products, module products, and data. The organization of these computer software units is given in Appendix B.

Procedures for Changing Baseline. This section should describe the procedures for making controlled changes to the development baseline. The following provides an example of this type of information:

> Changes to each software product baseline are made according to the change process.
>   Change Tracking. Changes to each software product baseline are tracked using an automated CER and SCR tracking system. These systems report current status for all CERs and SCRs for a specified version and product.
>   Change Release. The PL provides reports containing the following information for review at each CCB meeting for:
>
>   a. Number of new CERs
>   b. Type
>   c. Priority
>   d. Source (Internal, Field, FOT&E)
>   e. Action required
>   f. Number of CERs in each status
>
>   Each development team Project Lead, or designated Change Request Coordinator (CRC), is responsible for keeping the information in the CER tracking system up to date so the information is available to the PL and PM prior to CCB meetings.

Procedures for Processing Change Requests and Approvals. This section should describe the procedures for processing changes requests and the associated approval process. An example of a change request process is provided in Appendix C, Software Change Request Procedures, of this book.

Organizations Assigned Responsibilities for Change Control. This section should describe all organizations associated with the change control process and their associated responsibilities.

Change Control Boards (CCBs). This section should describe the role of the CCB and refer to any associated documentation (e.g., CCB charter).

Interfaces. This section should describe any required software interfaces and point to the source of their SCM documentation.

Level of Control. This section should describe the levels of control and implementing authorities. The following is an example:

> Evaluating Changes. Changes to existing software baselines are evaluated by the software product's associated CCB.
> Approving or Disapproving Changes. Changes to existing software baselines are either approved or disapproved by the associated project CCB.
> Implementing Changes. Approved changes are incorporated into the product baseline by the development team as described by the change control process documented in Appendix A.

Document Revisions. This section should describe SCM plan maintenance information. All activities and responsibilities necessary to ensure continued SCM planning during the life cycle of the project should be described. The Plan should describe:

    a. Who is responsible for monitoring the plan
    b. How frequently updates are to be performed
    c. How changes to the plan are to be evaluated and approved
    d. How changes to the plan are to be made and communicated

Automated Tools Used to Perform Change Control. This section should identify all software tools, techniques, equipment, personnel, and training necessary for the implementation of SCM activities. SCM can be performed by a combination of software tools and manual procedures. For each software tool, whether developed within the project or brought in from outside the project, the plan should describe or reference its functions and shall identify all configuration controls to be placed on the tool.

Configuration Status Accounting. Configuration status accounting activities record and report the status of a project's CIs. The SCM plan should describe what will be tracked and reported, describe the types of reporting and their frequency, and how the information will be processed and controlled.

If an automated system is used for any status accounting activity, its function should be described or referenced. The following minimum data elements should be tracked and reported for each CI: its initial approved version, the status of requested changes, and the implementation status of approved changes. The level of detail and specific data required may vary according to the information needs of the project and the customer.

Storage, Handling, and Release of Project Media. This section should describe the storage, handling, and release of SCM products. The following is provided as an example:

> The initial approved version of a [Project Abbreviation] software product is separately maintained by the PL until the approval of subsequent version of the product. Upon approval of

the subsequent version, the initial version is held in the project SCM library. This library is under configuration control.

Reporting. This section should describe the format, frequency, and process for reporting SCM product baseline status. Providing configuration items to testing may require a test item transmittal report. This test item report should identify the person responsible for each item, the physical location of the item, and item status. The following is an example of a status report of requested changes:

> The status report of requested changes is available to the PM at any time. A schedule status report may be provided to the CCB weekly and reported to [Company Name] senior management on a monthly basis. Please refer to the [Project Abbreviation] Software Development Plan, Section 6, for a status report schedule example.

Release Process. This section should describe the release process, including all required approvals. A summary chart is sometimes used to describe the release process effectively.

Document Status Accounting. This section should describe how changes to the SCM plan are managed and reported.

Change Management Status Accounting. This section should describe the format, frequency, and process for reporting SCM change management status.

Configuration Audits and Reviews. This section should provide information in support of required configuration audits and reviews. The following is provided as an example:

> Each team PL reviews the software baseline semiannually. The [Project Abbreviation] PM and [Project Abbreviation] SCM review the baseline of each software product annually. This baseline review consists of checking for CER implementation. The CERs are randomly selected. Discrepancies will be annotated and reported to the PL. Corrective action is taken if required.
> The [Project Abbreviation] CCB reviews the SCM process annually. Randomly selected problem reports are identified and walked through the change process. Discrepancies are annotated and reported to the [Project Abbreviation] PM and [Project Abbreviation] SCM. Corrective action is taken if required.

A software baseline, the associated status reporting, and associated documentation for each software product is available for SQA review at all times.

CM Milestones. This section should provide a description of the minimum set of SCM related project milestones that are acceptable for compliance. These milestones should be reflected in the software project management plan schedule and resource allocation.

Training. This section can refer to an independent training plan. However, this training plan must include information relating specifically to SCM training. If an independent plan does not exist, then information regarding the type and frequency of SCM training should be identified here.

Subcontractor/Vendor Support. This section should describe the SCM activities required by subcontractors. The following information is provided as a supporting example:

[Project Abbreviation] subcontractor support will be identified by the [Project Abbreviation] Program Manager. Subcontractor support will be required to follow the process defined in this document. If a wavier is requested, the subcontractor must provide evidence of comparable configuration management procedures. These procedures will follow the same audit and control procedures described in Section xxx.

APPENDIX A. Software Change Request Procedures

Refer to Appendix C, Software Change Request Procedures, of this book for example software change request procedures.

APPENDIX B. [PROGRAM ABBREVIATION] Software Organization

B.1 [PROGRAM ABBREVIATION] Organization

[PROGRAM ABBREVIATION] software elements can be categorized into three groups: module products, data products, and end-user products.

Module products and data products are used by software developers and not the users, and are components of [PROGRAM ABBREVIATION].

Module Products. Module products are software libraries that perform a specific set of functions. Module products provide reusable code, ensuring consistency in function performance, eliminating duplication of effort in developing like functions, and reducing the amount of code. Module products are used in our end-user products and may be available for use by other companies/agencies.

Data Products. Data products are database data files. Data products allow a single development, maintenance, and testing source and provide consistent data and format. Data products are included as part of our end-user products and may be available for use by other companies/agencies. Data products may be sent to users as a product itself.

End-user Products. End-user products are executable programs for users.

Configuration management of module products is accomplished according to this document. Each module product has an individual baseline and version numbers. Module product managers specify the version of the module product to be used in end-user products.

B.1.1 [PROGRAM ABBREVIATION] Module Products

A PVCS directory structure for all module products exists for each software product version released.

B.1.2 [PROGRAM ABBREVIATION] Data Products

B.1.3 [PROGRAM ABBREVIATION] Deliverables

B.1.4 [PROGRAM ABBREVIATION] Support Software

SUPPLIER AGREEMENT MANAGEMENT

This PA focuses on the processes supporting the acquisition of products from suppliers for which there exists a formal agreement. The Supplier Agreement Management process area addresses what is required in support of effective acquisition of work produced by suppliers external to an organization. A supplier agreement is established and maintained and is used to manage the supplier. This PA requires that all progress and performance be monitored and reported.

The CMMI®-SW is more rigorous in its requirements in support of acquisition plan maintenance. IEEE Std 1062, IEEE Recommended Practice for Software Acquisition

[17], provides guidance in support of the documentation required to support the CMMI®-SW PA for Supplier Agreement Management. Appendix B of this IEEE Software Engineering standard provides a suggested document format with detailed guideline support for each recommended document section.

IEEE Std 12207.0, Standard for Information Technology—Software Life Cycle Processes [39], describes 17 processes spanning the entire life cycle of a software product or service. Even if an organization's processes were defined using other sources, the standard is useful in characterizing the essential characteristics of these software processes and should be considered prior to the implementation of process improvement activities. Referencing this standard and reviewing what is required for each of these primary process areas can provide additional guidance in support of the activities associated with process supplier agreement management. It is important that IEEE 12207 be considered prior to the implementation of any process improvement activities associated with supplier agreement management.

CMMI®-SW Goals

IEEE Std 1062 and 12207 support the analysis and definition of the processes supporting the software acquisition life cycle. Complete guidance for the CMMI®-SW PA Supplier Agreement Management is provided when used in conjunction with the IEEE standard supporting software project planning, IEEE Std 1058.

SG 1. Establish Supplier Agreements
Agreements with the suppliers are established and maintained.

There is no obvious IEEE requirement to have a software development plan for each subcontractor; this is the CMM® Level 2 basis for monitoring the progress of the supplier. If you dig, you find the requirement in Section 4.7.7, Subcontractor Management Plans, of the 1058 Standard. This section states:

> The SPMP shall contain plans for selecting and managing any subcontractors that may contribute work products to the software project. The criteria for selecting subcontractors shall be specified and the management plan for each subcontract shall be generated using an adapted version of this standard. Plans should include the monitoring of technical progress, schedule and budget control, and product acceptance criteria, and risk management procedures shall be included in each subcontractor plan. Additional topics should be added as needed to ensure successful completion of the subcontract. A reference to the official subcontract and prime contractor/subcontractor points of contact shall be specified. [15]

SG2. Satisfy Supplier Agreements
Agreements with the suppliers are satisfied by both the project and the supplier.

IEEE Std 1058-1998, IEEE Standard for Software Project Management Plans, Section 4.7.7, Subcontractor Management Plans, and IEEE Std 1062-1998, IEEE Recommended Practice for Software Acquisition, each provide valuable insight into the items required for effective subcontractor elicitation, evaluation, acquisition, and management. This section states generally what is required and states that an adapted version of 1058 should be used for subcontracted software management plans. However, more information would benefit the standards users.

GG2. Institutionalize a Managed Process
The process is institutionalized as a managed process.

The CMMI® requires the definition of the acquisition process and the requirements for each product to be acquired. It also incorporates the requirement to review of candidate COTS products to ensure they satisfy the specified requirements covered in the supplier agreement. IEEE Std 1062 can be used as a resource in support of these requirements. This standard describes the software acquisition life cycle and supporting processes in detail. This standard includes a checklist that may be used by organizations when establishing a software acquisition process and detailed guidelines for supporting acquisition planning. IEEE Std 12207.0 provides information in support of associated process requirements and activities. This standard also describes the objectives associated with the acquisition process. Table 5-10 describes the specific and generic goals and practices, typical work products, and commonly associated plans or artifacts in support of the PA.

Software Acquisition Plan

A clear description of how adequately IEEE Std 1062 and IEEE Std 12207 support the CMMI®-SW for Supplier Agreement Management can be seen in Recommendations for Software Acquisition, in Appendix C of this book. IEEE Std 12207 and IEEE Std 1062 support each step with prescriptive detail. The modification of the recommended software acquisition plan (SAP) table of contents to support the goals of the CMMI® more directly is shown in Table 5-11. Appendix C, Software Process Work Products, provides additional information relating to the work products associated with software acquisition activities. These include a discussion of the nine steps of acquisition, an organizational acquisition strategy checklist, supplier evaluation criteria, and supplier performance standards.

Software Acquisition Plan Document Guidance

The following provides section-by-section guidance in support of the creation of a SAP. The SAP should be considered to be a living document and should change to reflect any process improvement activity. This guidance should be used to help define a software acquisition process and should reflect the actual processes and procedures of the implementing organization. Additional information is provided in the document template, *Software Acquisition Plan.doc,* which is located on the CD-ROM accompanying this book.

Introduction. The SAP should describe the specific purpose, goals, and scope of the software acquisition effort. All associated requirements and plans should be identified. The type of contract and support concept should be identified.

References. This section should all documents all supporting documents supplementing or implementing the SAP, including other plans, processes, or task descriptions that elaborate details of this plan.

Definitions and Acronyms. The SAP should define or reference all terms unique to the development and understanding of the SAP. All abbreviations and notations used in the SAP should also be described.

Table 5-10. Supplier agreement management goals and practices [61]

Specific and generic goals/practices and typical work products	Plan or artifact
SG1. Establish Supplier Agreements	
SP 1.1. Determine acquisition type	
List of all acquisition types that will be used for all products and product components to be acquired	Software Acquisition plan or statement of work (SOW)
SP 1.2. Select suppliers	
List of candidate supplier	Software acquisition plan
Preferred supplier list	Software acquisition plan
Rationale for selection of suppliers	Software acquisition plan
Advantages and disadvantages of candidate suppliers	Software acquisition plan
Evaluation criteria	Software acquisition plan
Solicitation materials and requirements	Software acquisition plan
SP 1.3. Establish supplier agreements	
Statements of work	Software acquisition plan or software project management plan
Contracts	Software acquisition plan or software project management plan
Memoranda of agreement	Software acquisition plan or software project management plan
Licensing agreement	Software acquisition plan or software project management plan
SG2. Satisfy Supplier Agreements	
SP 2.1. Review COTS products	
Trade studies	Software acquisition plan or software project management plan
Price lists	Software acquisition plan or software project management plan
Evaluation criteria	Software acquisition plan or software project management plan
Supplier performance reports	Software acquisition plan or software project management plan
Reviews of COTS products	Software acquisition plan or software project management plan
SP 2.2. Execute the supplier agreement	
Progress reports and performance measures	Software acquisition plan or software project management plan
Supplier review materials and reports	Software acquisition plan or software project management plan
Action items tracked to closure	Software acquisition plan or software project management plan
Documentation of deliverables	Software acquisition plan or software project management plan
SP 2.3. Accept the acquired product	
Acceptance test procedures	Software acquisition plan or software project management plan

Table 5-10. *Continued*

Specific and generic goals/practices and typical work products	Plan or artifact
SG2. Satisfy Supplier Agreements (*cont.*)	
SP 2.3. Accept the acquired product (*cont.*)	
Acceptance test results	Unit test report
Discrepancy reports or corrective action plans	Software acquisition plan or software project management plan
SP 2.4. Transition Products	
Transition plans	Software acquisition plan
Training reports	Training plan or software acquisition plan
Support and maintenance reports	Software acquisition plan or transition plan
GG2. Institutionalize a Managed Process	
GP 2.1. Establish an organizational policy	Organizational policy supporting supplier agreement management
GP 2.2. Plan the process	Software acquisition plan and/or contracts regulatory documentation
GP 2.3. Provide resources	
Identification of resources used in support of supplier agreement management	Software acquisition plan
GP 2.4. Assign responsibility	Software acquisition plan
GP 2.5. Train people	Software acquisition plan or training plan
GP 2.6. Manage configurations	
Description of how work products and all associated artifacts in support of supplier agreement management are placed under configuration management	Configuration management plan
GP 2.7. Identify relevant stakeholders	
Supplier lists, reviews, and agreements are typical artifacts	Software acquisition plan
GP 2.8. Monitor and control the process	
How project volatility will be reported	Software measurement and metrics plan, requirements management plan, unit test plan, or software project management plan
GP 2.9. Objectively evaluate adherence	Software quality assurance plan or separate supplier agreement
GP 2.10. Review status with higher-level management	Software acquisition plan
GG3. Institutionalize a Defined Process	
GP 3.1. Establish a defined process	
Establish and maintain a description of the supplier agreement management process	Software acquisition plan
GP 3.2. Collect improvement information	Requirements management plan, software measurement and metrics plan, or software acquisition plan

Table 5-11. Software acquisition plan document outline [17]

Title Page
Revision Page
Table of Contents
1. Introduction
2. References
3. Definitions and Acronyms
4. Software Acquisition Overview
    4.1 Organization
    4.2 Schedule
    4.3 Resource summary
    4.4 Responsibilities
    4.5 Tools, techniques, and methods
5. Software Acquisition Process
    5.1 Planning organizational strategy
    5.2 Implementing the organization's process
    5.3 Determining the software requirements
    5.4 Identifying potential suppliers
    5.5 Preparing contract documents
    5.6 Evaluating proposals and selecting the suppliers
    5.7 Managing supplier performance
    5.8 Accepting the software
    5.9 Using the software
6. Software Acquisition Reporting Requirements
7. Software Acquisition Management Requirements
    7.1 Anomaly resolution and reporting
    7.2 Deviation policy
    7.3 Control procedures
    7.4 Standards, practices, and conventions
    7.5 Performance tracking
    7.6 Quality control of plan
8. Software Acquisition Documentation Requirements

Software Acquisition Overview. The section should describe the acquiring organization, acquisition schedule, and associated resources. All responsibilities, tools, techniques, and methods necessary to perform the software acquisition process should also be described.

Organization. This subsection of the SAP should describe the organization of the acquisition effort. This description of the organization should include the approval hierarchy and points of contact.

Schedule. This subsection should describe the schedule of the acquisition and include all milestones. If this schedule is included as part of the acquiring organization's software project management plan, this plan may be referenced.

Resource Summary. This subsection should describe all staffing, facilities, tools, finances, and any special procedural requirements that are needed in support of the software acquisition effort.

Responsibilities. This subsection should provide an overview of acquisition responsibilities.

Tools, Techniques, and Methods. This subsection should describe all documentation, tools, techniques, methods, and environments to be used during the acquisition process. Acquisition, training, support, and qualification information for each should be provided. The SAP should document the metrics to be used by the acquisition process and should describe how these metrics support the acquisition process.

Software Acquisition Process. This section should identify all actions to be performed for each of the software acquisition steps shown:

1. Planning organizational strategy
2. Implementing the organization's process
3. Determining the software requirements
4. Identifying potential suppliers
5. Preparing contract documents
6. Evaluating proposals and selecting the suppliers
7. Managing supplier performance
8. Accepting the software
9. Using the software

For further descriptions, refer to the table in Appendix C (Software Process Work Products) entitled Recommendations for Software Acquisition. Any additional information required in support of the software acquisition process should be included as deemed necessary.

Software Acquisition Reporting Requirements. This section should describe all reporting requirements. It should include a description of the reporting in support of acquisition status and risk.

Software Acquisition Management Requirements. This section should describe all procedures and processes in support of anomaly resolution and reporting. References should be provided to other plans describing the management of the software acquisition process. The subsections in this section are:

Anomaly Resolution and Reporting. This subsection describes the method of reporting and resolving anomalies, including all reporting and resolution criteria.

Deviation Policy. This subsection should describe all procedures and forms used if deviation is required. All deviation approval authorities should also be identified.

Control Procedures. This subsection should describe all procedures and processes supporting the configuration, protection, and storage of associated software products.

Standards, Practices, and Conventions. This subsection of the plan should describe all standards, practices, and conventions used in the development of this plan or in support of the acquisition process.

Performance Tracking. This subsection should describe the processes supporting performance monitoring. All items tracked should be identified. All reporting procedures should also be described, including reporting format.

Quality Control of the Plan. This subsection should describe the processes supporting the development and maintenance of this plan. Other project plans may be referenced in support of this section (e.g., software quality assurance plan).

Software Acquisition Documentation Requirements. All documentation requirements required in support of the acquisition process should be described here. An appendix providing examples of all required document formats could be included.

MEASUREMENT AND ANALYSIS

IEEE Std 1058, Standard for Software Project Management Plans, points to the requirement of a measurement and metrics plan. This plan must identify the specific measurement and analysis activity to be provided in support of a software effort. The metrics plan should also include information about how the metrics are stored in any associated metrics database, which metrics are reported to senior management above the project level in order for senior management to perform reviews of the project, and a description of procedures that are included in the metrics process for ensuring that metrics are not just collected, analyzed, and forgotten, but actions are taken based on the results of analyzing the metrics. As previously described, there are a number of IEEE Standards supporting this PA, in particular, the IEEE Standard Classification for Software Anomalies (1044) and Standard for Productivity Metrics (1045).

However, there are areas of weak support. The requirement to describe how the measurement and analysis data are stored and maintained should be addressed either in the measurement and metrics plan, or as a cross reference to the project SCM plan. The identification of relevant stakeholders should also either be addressed in the measurement and metrics plan, or as a cross reference to the project management plan. Responsibility and training are areas that should also be directly addressed during the documentation of the processes in support of this PA.

IEEE Std 12207.0, Standard for Information Technology—Software Life Cycle Processes [39], describes 17 processes spanning the entire life cycle of a software product or service. Even if an organization's processes were defined using other sources, the standard is useful in characterizing the essential characteristics of these software processes and should be considered prior to the implementation of process improvement activities. Referencing this standard and reviewing what is required for each of these primary process areas can provide additional guidance in support of the activities associated with measurement and analysis. It is important that IEEE 12207 be considered prior to the implementation of any process improvement activities associated with measurement and analysis.

CMMI®-SW Goals

Several IEEE software engineering standards support the definition and implementation of software measurement and metrics. These standards support a uniform approach to the definition and measurement of productivity. These goals are primarily supported by IEEE Std 1044, Standard Classification for Software Anomalies; IEEE Std 982.1, Standard

Dictionary of Measures to Produce Reliable Software [7]; IEEE Std 1045, Standard for Software Productivity Metrics [14]; IEEE Std 1061, Software Quality Metrics Methodology [16]; IEEE Std 12207.0, Guide to Lifecycle Processes [39], and ISO/IEC 15939:2001, Information Technology—Software Engineering—Software Measurement Process [88]. These standards provide a common language and framework in support of software measurement and analysis activities. The information provided by these plans support the goals of the CMMI®-SW Measurement and Analysis PA as long as they are used in conjunction with the other IEEE standards supporting project planning, software configuration management, and software quality assurance.

SG1. Align Measurement and Analysis Activities
Measurement objectives and activities are aligned with identified information needs and objectives.

The IEEE Standard for Developing Software Life Cycle Processes (SLCP), IEEE Std 1074 [19], provides for the definition and maintenance of the processes that govern a software project. This standard applies to the management and support activities throughout the entire software lifecycle. This standard also provides guidance in all aspects of the software life cycle from concept exploration through retirement.

Annex A of Standard 1074 describes all the required activities in support of SLCP development, including the definition and evaluation of project metrics. Annex B of this standard provides an example of SCLP development. Annex C provides an informative mapping template that can be used as a checklist to identify and track required project deliverable.

ISO/IEC 15939 has a Measurement Information Model that describes how the relevant attributes are quantified and converted to indicators that provide a basis for decision making. Please refer to the example provided in Appendix C of this book.

SG2. Provide Measurement Results
Measurement results that address identified information needs and objectives are provided.

IEEE Std 982.1, IEEE Standard IEEE Standard Dictionary of Measures to Produce Reliable Software [7], provides a set of software reliability measures that can be applied to the software product as well as to the supporting development processes. This standard provides measures that can be applied early in the development process. This standard provides a common set of measures and is designed to assist management in product management and oversight activities. The standard describes and supports measures for both software product and process.

CG2. Institutionalize a Managed Process
The process is institutionalized as a managed process.

IEEE Std 1074 and the IEEE 12007 series can help process architects develop the SLCPs that are required in support of a specific software project. IEEE Std 1074 does not support the development of organizational software life cycle processes. If support of organizational process development is needed, organizations should take advantage of the information provided in the IEEE 12207 series [39]: Software Life Cycle Processes (12207.0),

Table 5-12. Measurement and analysis goals and practices [61]

Specific and generic goals/practices and typical work products	Plan or artifact
SG1. Align Measurement and Analysis Activities	
SP 1.1. Establish measurement objectives	
Measurement objectives	Software project management plan, software measurement and metrics plan, requirements management plan, or other supporting project documentation
SP 1.2. Specify measures	
Specifications of base and derived measures	Software measurement and metrics plan, measurement information model
SP 1.3. Specify data collection and storage procedures	
Data collection storage procedures	Software measurement and metrics plan or software project management plan
Data collection tools	
SP 1.4. Specify analysis procedures	
Analysis specification and procedures	Software measurement and metrics plan or software project management plan or measurement information model
Data analysis tools	Software measurement and metrics plan or software project management plan
SG2. Provide Measurement Results	
SP 2.1. Collect measurement data	
Base and derived measurement data sets	Software measurement and metrics plan or software project management plan
Results of data integrity tests	Software measurement and metrics plan or software project management plan
SP 2.2. Analyze measurement data	
Analysis results and draft reports	Software measurement and metrics plan or software project management plan
SP 2.3. Store data and results	
Stored data inventory	Configuration management plan
SP 2.4. Communicate results	
Delivered reports and related analysis results	Software measurement and metrics plan or software project management plan
Contextual information or guidance to aid in the interpretation of analysis results	Software measurement and metrics plan or software project management plan
GG2. Institutionalize a Managed Process	
GP 2.1. Establish an organizational policy	Organizational policy that establishes expectations for measurement objectives and activities
GP 2.2. Plan the process	Software measurement and metrics plan or software project management plan

Table 5-12. *Continued*

Specific and generic goals/practices and typical work products	Plan or artifact
GG2. Institutionalize a Managed Process (*cont.*)	
GP 2.3. Provide resources	
Identification of resources used in support of measurement and analysis activities	Software measurement and metrics plan or software project management plan
GP 2.4. Assign responsibility	Software project management plan, metrics plan, or organizational policy
GP 2.5. Train people	Software project management plan or training plan
GP 2.6. Manage configurations	Configuration management plan
GP 2.7. Identify relevant stakeholders	
Describe measurement analysis feedback mechanisms	Software project management plan, metrics plan, or organizational policy
GP 2.8. Monitor and control the process	
How project measurement and analysis activity will be reported	Software measurement and metrics plan or software project management plan
GP 2.9. Objectively evaluate adherence	Software quality assurance plan
GP 2.10. Review status with higher-level management	Software project management plan
GG3. Institutionalize a Defined Process	
GP 3.1. Establish a defined process	
Establish and maintain a description of the measurement and analysis process	Software measurement and metrics plan
GP 3.2. Collect improvement information	Requirements management plan, software measurement and metrics plan, or software project management plan

Life Cycle Data (12207.1), and Implementation Considerations (12207.2). Table 5-12 describes the specific and generic goals and practices, typical work products, and commonly associated plans or artifacts in support of the PA.

IEEE/EIA 12207.1 lists all of the data produced by the processes of IEEE/EIA 12207.0 and provides a description of that data. IEEE Std 1074 provides a set of process building blocks that can be applied to define project processes based on the requirements of 12207 or other life cycle process standards. [19]

The Measurement and Analysis (MA) PA defines the processes supporting the development, maintenance, and implementation of software project measurement activities. These activities support objective planning and estimating for all process areas tracking actual performance against initial estimates. The results from MA activities can be used in making informed decisions and taking appropriate corrective actions. Remember that collecting metrics consumes project resources. Start small and concentrate on the metrics that will most effectively promote change.

In addition, the software project management plan should provide information about estimating project attributes, and the monitoring of the project. The configuration management plan should provide additional information about the management of measurement work products. Lastly, the requirements management plan should provide information regarding traceability.

Software Measurement and Metrics Plan

The following provides a suggested format for a project level measurement and metrics* plan. This plan should contain a description of all measurement and analysis used in support of an identified software effort. The modification of the recommended software measurement and metrics plan table of contents to support the goals of the CMMI® more directly is shown in Table 5-13. Additional information in support of related work products is provided in Appendix C, Software Process Work Products. These include a list of measures for reliable software and the measurement information model as described in ISO/IEC 15939.

Software Measurement and Metrics Plan Document Guidance

The following provides section-by-section guidance in support of the creation of a software measurement and metrics plan (MMP). The MMP should be considered to be a living document. The MMP should change to reflect any process improvement or changes in procedures relating to software measurement and metrics activities. This guidance should be used to help define a measurement and metrics process and should reflect the actual processes and procedures of the implementing organization. Additional information is provided in the document template, *Software Measurement and Metrics Plan.doc,* which is located on the CD-ROM accompanying this book.

INTRODUCTION

An introduction is not required, but can be used to state the goals for measurement and metrics activities. The following information is provided as an example introduction:

> The goal of the [Project Name] ([Project Abbreviation]) software development effort is to [Project Goal]. This software will provide a method for [Customer Name] to efficiently meet their customer's requirements.
>
> This Software Measurement and Metrics (SMM) Plan has been developed to ensure that [Project Abbreviation] software products conform to established technical and contractual requirements.

---

*Metrics is a problematic term because it has no counterpart in generally accepted metrology. Generally, it is used in software engineering to mean a measurement coupled with a judgmental threshold of acceptability.

Table 5-13. Software measurement and metrics plan document outline

Title Page
Revision Page
Table of Contents
1. Introduction
    1.1 Purpose
    1.2 Scope
    1.3 Definitions, Acronyms, and Abbreviations
    1.4 References
2. Metrics Process Management
    2.1 Responsibilities
    2.2 Life Cycle Reporting
    2.3 General Information
3. Measurement and Metrics

Purpose. This section should describe the objectives of measurement and analysis activities and how these activities support the software process. An overview of all measures, data collection and storage mechanisms, analysis techniques, and reporting and feedback mechanisms should be provided. An example is provided:

> The purpose of this SMM Plan is to establish, document, and maintain consistent methods for measuring the quality of all software products developed and maintained by the [Project Abbreviation] development team. This plan defines the minimum standardized set for information gathering over the software life cycle. These metrics serve as software and system measures and indicators that critical technical characteristics and operational issues have been achieved.

Scope. This section should provide a brief description of scope, including the identification of all associated projects and a description of the metric methodology employed. An example is provided below:

> The measures to be utilized by [Project Abbreviation] are described in detail in Section 3. These metrics deemed to be the most feasible and cost effective to initially implement are described in Table 1-2 below.
>
> The metrics employed by [Project Abbreviation] fall into three general categories as shown in Table 1-2. Management metrics deal with contracting, programmatic, and overall management issues. Requirements metrics pertain to the specification, translation, and volatility of requirements. Quality metrics deal with testing and other software technical characteristics.
>
> Additional metrics were evaluated for [Project Abbreviation] Program implementation. Those metrics reviewed and deemed "not applicable" are noted as NA.
>
> The specific products developed and maintained by the [Project Abbreviation] development team may be found in the [Project Abbreviation] Software Configuration Management (SCM) Plan.

Table 1-2. Program metrics

Metric	Producer	Objective	Measurement
		Management Category	
Manpower	Program Manager	Track actual and estimated man-hours usage	Hours used vs. estimated
Development Progress	Project Manager	Track computer software units (CSU) planned vs. coded	Number of coded CSUs
Cost	NA	Track S/W expenditures	NA
Schedule	Project Manager	Track schedule adherence	Milestone/event slippage
Computer Resource Utilization	NA	Track planned and actual resource use	NA
Software Engineering Environment (This metric is not covered in section 3)	[Company Name] EPG	Quantify developer S/W engineering environment maturity	Computed # of process areas (PA)s
		Requirements Category	
Requirements Traceability	Project Manager	Track requirements to code	% of requirements traced throughout program life cycle
Requirements Stability	Project Manager	Track changes to requirements	# and % of requirements changed/added per project
		Quality Category	
Design Stability	Project Manager	Track design changes	Stability index
Complexity	Not implemented due to manpower constraints	Assess code quality	Complexity indices
Breadth of Testing	Project Manager	Track testing of requirements	% requirements tested, and % reqmnts passed
Depth of Testing	Project Manager	Track testing of code	Degree of code testing
Fault Profiles	Project Manager	Track open vs. closed anomalies	# and types of faults, average open age

Definitions, Acronyms, and Abbreviations. This section should include all definitions and acronyms used during the development of this plan.

References. This section should include a list of all references used during the development of this plan.

Metrics Process Management. This section should describe the process for the definition, selection, collection, and reporting of measurement and metrics. Using a diagram to help describe the process flow is often helpful. The following provides an example of information typical of this section:

> The [Project Abbreviation] Program Manager, the [Project Abbreviation] Project Manager, and the [Company Name] Engineering Process Group (EPG) will work together to develop, maintain, and implement an effective software measurements and metrics program. The results of all metrics should be reported at Program Management Reviews (PMRs), Configuration Control Board (CCB) reviews and Pre-Release/validation reviews. Figure 2-1 provides an overview of the [Project Abbreviation] metrics collection and reporting process.

Responsibilities. This section should describe all responsibilities related to the measurement and metrics process. Subsections 2.1.1 and 2.1.2 provide an example of this information:

> Program Manager. The [Project Abbreviation] Program Manager is ultimately responsible for ensuring that Software Measurement and Metrics (SMM) procedures are developed, im-

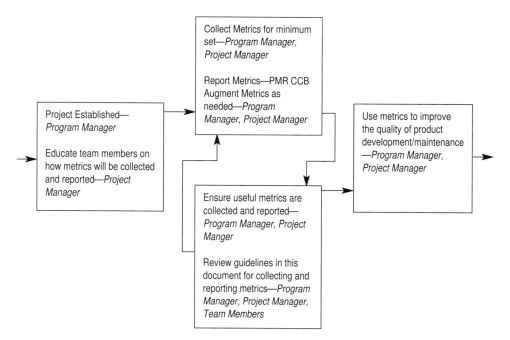

Figure 2-1. [Project Abbrev.] Software Quality Measurement and Metrics Process

plemented, and maintained. In this capacity, the Program Manager will work with the [Company Name] EPG and applicable [Project Abbreviation] CCBs in establishing standards and implementing EPG recommended metrics.

The Program Manager is ultimately responsible for ensuring that the [Project Abbreviation] Project Manager and the associated [Project Abbreviation] team is trained in SMM policies and procedures that apply to their function on the team. This will consist of initial training for new members and refresher training, as required, due to process or personnel changes.

Project Manager. The [Project Abbreviation] Project Manager shall provide direct oversight of SMM activities associated with project software development, modification, or maintenance tasks assigned within project functional areas. (Refer to Table 1-2.)

The Project Manager is responsible for ensuring that project SMM are collected and reported to the Program Manager. (Refer to Figure 2-1.)

Life Cycle Reporting. This section should describe all minimum reporting requirements. The following provides an example:

Figure 2-2 shows the applicability of the minimum metric set over the software life cycle. These metrics can provide valuable insight into a program, especially with regard to demonstrated results and readiness for test. The results of all metrics should be reported at Program Reviews, Staff (Project) Reviews, and all Validation/Test Reviews. Figure 2-2 also identifies the metrics that should be reported at each major decision milestone. Any other metric that indicates the potential for serious problems should also be reported at indicated milestones.

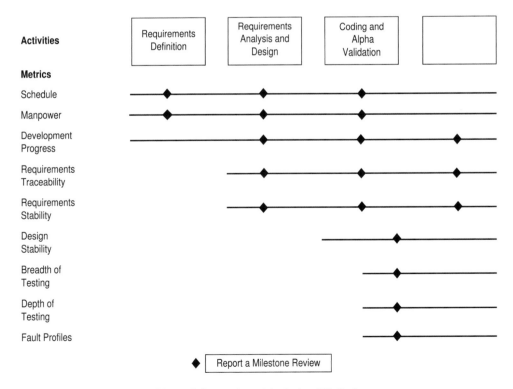

Figure 2-2. Metrics and the Project Life Cycle

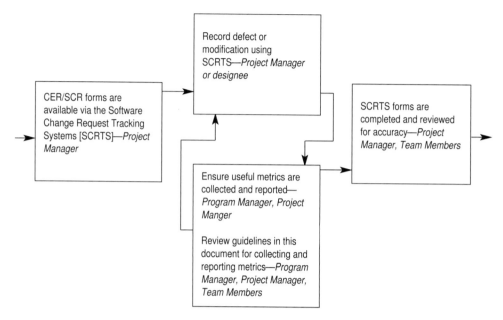

Figure 2-3. Collecting Metrics for Each Activity

The applicable time periods for data collection and analysis are provided with each metric description. For most metrics, data continues to be collected after the system is fielded. The metrics collected will be used as program maturity status indicators. They will be used to portray trends over time, rather than single point values. When a metrics base is established, trends will be compared with data from similar systems.

General Information. This section should provide any additional relevant information not previously provided. The following is an example:

The graphical displays shown in this plan (see Figure 2-3) are provided for illustrative purposes. There may be other ways of processing and displaying the data that are more appropriate for a specific [Project Abbreviation] event. Specific project reports will be archived by reporting date by the Program Manager or designee.

Measurement and Metrics. This section (the rest of the plan) should provide a detailed description of all measurement collection and metrics calculation. This should also include data requirements, the frequency and type of reporting, interpretation recommendations, and examples when possible. It is often helpful to break this down by categories (e.g., management, requirements, quality). Detailed information should be provided in support of all metrics defined within a given category. Example measures are provided in Appendix C, Example Measurements, of this book.

6

CMMI-SW LEVEL 3

REQUIREMENTS DEVELOPMENT

Requirements Development (RD) analyzes and produces customer, product, and product-component requirements. The RD process area (PA) is part of the Engineer Process Category, which contains six PAs: Requirements Management, Technical Solution, Product Integration, Verification, and Validation. RD interfaces directly with the other five PAs. The RD PA has inputs from the customer and supplies requirements to both the Technical Solution and Product Integration PAs. The RD PA operates under the management of other PAs: Project Planning, Project Monitoring and Control, and Requirements Management.

The RD PA identifies customer needs and translates these into product requirements. This set of product requirements is analyzed to produce a high-level conceptual solution. This set of requirements is then allocated to a set of product-component requirements. Other requirements that help define the product are derived and allocated to product components, with inputs from the Technical Solution, Verification, Validation, and Product Integration PAs. This set of product and product-component requirements clearly describes the product's features, performance, key design features, V&V requirements, and so on, in terms the developer and tester understand and use. RD is a key PA as most product defects have been traced to RD errors.

The purpose of RD is to produce and analyze customer, product, and product-component requirements. These requirements address the relevant stakeholder needs throughout the product life-cycle activities and the various product attributes. Requirements also address constraints caused by the selection of design solutions.

Software either covers the entire project or is embedded within a system of interest (e.g., aligned with Hardware requirements). Both software and system requirement templates will be provided here. Additional information in support of requirements change management is provided in Appendix C.

RD results in several typical work products or artifacts, some of which are expected in other PAs. IEEE Standards 1362, Guide for Information Technology Concept of Operations [28]; 830, Recommended Practice for Software Requirements Specifications [6]; and 1233, Guide for Developing System Requirements Specifications [25] provide detailed guidance in support of the development of customer requirements and software requirements documents.

IEEE Std 12207.0, Standard for Information Technology—Software Life Cycle Processes [39], describes 17 processes spanning the entire life cycle of a software product or service. Even if an organization's processes were defined using other sources, the standard is useful in characterizing the essential characteristics of these software processes and should be considered prior to the implementation of process improvement activities. Referencing this standard and reviewing what is required for each of these primary process areas can provide additional guidance in support of the activities associated with process requirements development. It is important that IEEE 12207 be considered prior to the implementation of any process improvement activities associated with process requirements development.

CMMI®-SW Goals

The RD PA has three specific goals (SGs) and one generic goal (GG):

SG 1. Stakeholder needs, expectations, constraints, and interfaces are collected and translated into customer requirements. Specific support for this goal is provided by IEEE Standard 1362 Concept of Operations (ConOps) Document.

SG 2. Customer requirements are refined and elaborated to develop product and product-component requirements. Both IEEE Standard 830 Software Requirements Specification [6] and IEEE Standard 1233 System Requirements Specification [25] provide specific support for this goal.

SG 3. The requirements are analyzed and validated, and a definition of required functionality is developed. This goal results in a variety of typical work products. The best practice is to organize these work products into the overall SRS originating from SG 2, using configuration management (CM) controls.

GG3. The process is institutionalized as a defined process.

Maturity Level 3 requires that the RD PA be institutionalized as a defined process, so each project team can tailor their processes and document templates from the organization's set of standard processes according to the organization's tailoring guidelines. Then generic practice GP 2.7, Identify and Involve the Relevant Stakeholders as Planned will ensure that RD products will be complete. This PA is uses special CMMI® terminology for customer, stakeholder, and relevant stakeholder. Table 6-1 describes the specific and generic goals and practices, typical work products, and commonly associated plans or artifacts in support of the PA.

Software Requirements Specification

The information provided here in support of requirements management is designed to facilitate the definition of a software requirements specification. This information was developed using IEEE Std 830-1998, IEEE Recommended Practice for software require-

Table 6-1. Requirements development goals and practices [61]

Specific and generic goals/practices and typical work products	Plan or artifact
SG1. Develop Customer Requirements	
SP1.1. Elicit needs	Software requirements management plan or software project management plan
SP1.1-1. Collect stakeholder needs (continuous representation) Evidence of:	
Technology demonstrations	ConOps document
Interface control working groups	Software requirements specification or interface specification
Technical control working groups Interim project reviews Beta testing planning	
SP 1.2. Develop customer requirements	
Customer requirements	Software requirements specification, ConOps document
Customer constraints on the conduct of verification	Software project plan and procedures associated with walk-throughs and inspections
Customer constraints on the conduct of validation	Software test plans ConOps document
SG2. Develop Product Requirements	
SP 2.1. Establish product and product component requirements	
Derived requirements	Software requirements specification, system requirements specification
Product requirements	Software requirements specification, system requirements specification
Product component requirements	Software requirements specification, system requirements specification
SP 2.2. Allocate product component requirements	
Requirement allocation	Software requirements specification, system requirements specification
Design constraints	Software requirements specification, system requirements specification
Derived requirements	Software requirements specification, system requirements specification
Relationships among derived requirements	Software requirements specification, system requirements specification
SP 2.3. Identify interface requirements	
Interface requirements	Software requirements specification, system requirements specification, or interface specification
SG3. Analyze and Validate Requirements	
SP 3.1. Establish operational concept and scenarios	
Concept of operations document	ConOps document *(continued)*

Table 6-1. *Continued*

Specific and generic goals/practices and typical work products	Plan or artifact
SG3. Analyze and Validate Requirements (*cont.*)	
SP 3.1. Establish operational concept and scenarios (*cont.*)	
Product installation, operational, maintenance, and support concepts	ConOps document or transition plan
Disposal concepts	ConOps document
User cases	ConOps document, UML modeling
Timeline scenarios	ConOps document
New requirements	ConOps document
SP 3.2. Establish a definition of required functionality	
Functional architecture	Software requirements specification, system requirements specification
Activity diagrams	Software requirements specification, system requirements specification
Object-oriented analysis	Software requirements specification, system requirements specification
SP 3.3. Analyze requirements	
Requirements defects reports	Software requirements management plan
Proposed requirements changes to resolve defects	Configuration management report
Key requirements	Software requirements specification, System requirements specification
Technical performance measures	Software requirements specification, system requirements specification
SP 3.4. Analyze requirements to achieve balance	
Requirement related risk assessments	Software project management plan
SP 3.5. Validate requirements with comprehensive methods	
Record of analysis methods and results	Software test plan and results
SP3.5-1 Validate requirements (continuous representation)	
Results of requirements validation	Software test plan and results, requirements walk-through form
GG2. Institutionalize a Managed Process	Organizational policy
GP 2.1. Establish an organizational policy	
Establishes organizational expectations for collecting stakeholder needs, formulating product and product-component requirements, and analyzing and validating those requirements	
GP 2.2. Plan the process	Requirements management plan
GP 2.3. Provide resources	
Identification of resources used in support of requirements development activities	Requirements management plan, project plan

Table 6-1. *Continued*

Specific and generic goals/practices and typical work products	Plan or artifact
GG2. Institutionalize a Managed Process (*cont.*)	
GP 2.4. Assign responsibility	Software project management plan, requirements management plan
GP 2.5. Train people	Software project management plan, requirements management plan, or training plan
GP 2.6. Manage configurations	
Description of how work products and all associated artifacts in support of software requirements development activities are	Configuration management plan
GP 2.7. Identify relevant stakeholders	
Describe activities and feedback mechanisms	Software project management plan, requirements management plan
GP 2.8. Monitor and control the process	Requirements management plan
GP 2.9. Objectively evaluate adherence	Quality assurance plan reviews
GP 2.10. Review status with higher-level management	Software project management plan
GG3. Institutionalize a Defined Process	
GP 3.1. Establish a defined process	
Establish and maintain a description of the requirements development process	Requirements management plan
GP 3.2. Collect improvement information	Requirements management plan, software measurement and metrics plan, or software project management plan

ments specifications [6] that has been adapted to support CMMI® requirements. The modification of the recommended Software Requirements Specification table of contents to support the goals of the CMMI® more directly is shown in Table 6-2.

Software Requirements Specification Document Guidance

The following provides section-by-section guidance in support of the creation of a software requirements specification. This guidance should be used to help establish a requirements baseline and should reflect the actual processes and procedures of the implementing organization. Additional information is provided in the document template, *Software Requirements Specification.doc,* which is located on the CD-ROM accompanying this book.

Title Page. The title page should include the names and titles of all document approval authorities.

Revision Sheet. The revision sheet should provide the reader with a list of cumulative document revisions. The minimum information recorded should include the revision

Table 6-2. Software requirements
specification document outline

Title Page
Revision Page
Table of Contents
1. Introduction
    1.1 Software purpose
    1.2 Software scope
    1.3 Definitions, acronyms, and abbreviations
    1.4 References
    1.5 Software overview
2. Overall Description
    2.1 Product perspective
    2.2 Product functions
    2.3 Environmental conditions
    2.4 User characteristics
    2.5 External interfaces
    2.6 Constraints
        2.6.1 Safety
        2.6.2 Security
        2.6.3 Human Factors
    2.7 Assumptions and dependencies
3. Requirements Management
    3.1 Resources and funding
    3.2 Reporting procedures
    3.3 Training
4. Specific Requirements
5. Acceptance
6. Documentation
7. Maintenance
Appendixes
Traceability Matrix
Index

number, the document version number, the revision date, and a brief summary of changes.

Introduction. This section should explain the purpose and scope of the project software requirements specification (SRS), as well as provide clarification of definitions, acronyms, and references. This section should also provide an overview of the project.

Software Purpose. This subsection should explain the purpose for writing an SRS for this project and describe the intended audience for the SRS.

Software Scope. This subsection should identify the software products to be produced, by name; explain what the software products will, and, if necessary, will not do; and describe the application of the software being specified, including all relevant goals, objectives, and benefits from producing the software.

Definitions, Acronyms, and Abbreviations. This subsection should provide the definitions of all terms, acronyms, and abbreviations required to properly interpret the SRS.

Key References. This subsection should list all references used within the SRS.

Software Overview. This subsection should describe what the rest of the SRS contains as it relates to the software effort. It should also explain how the document is organized.

Overall Description. This section should describe the general factors that affect the product and its requirements.

Product Perspective. This subsection should put the product into perspective with other related products or projects. If the product to be produced from this SRS is totally independent, it should be clearly stated here. If the product to be produced from this SRS is part of a larger system, then this subsection should describe the functions of each component of the larger system or project and identify the interfaces between this product and the remainder of the system or project. This subsection should identify all principal external interfaces for this software product (Note: descriptions of the interfaces will be contained in another part of the SRS.) The product perspective requirements may be documented in the Product Packaging Information section in Appendix C.

Product Functions. This subsection should provide a summary of the functions to be performed by the software produced as a result of this SRS. Functions listed in this subsection should be organized in a way that will make it understandable to the intended audience of the SRS. (Note: this subsection is an overview; details of the specific requirements will be contained in Section 4 of the SRS.)

Environmental Conditions. This subsection should provide a summary of the environment in which the software must operate. (Note: this subsection is an overview; details of the specific requirements will be contained in Section 4 of the SRS.)

User Characteristics. This subsection should describe the general characteristics of the eventual users of the product that will affect the specific requirements. Eventual users of the product will include end-product customers, operators, maintainers, and systems people, as appropriate. For any users that impact the requirements, characteristics such as education, skill level, and experience levels will be documented within this subsection, as they impose constraints on the product.

External Interfaces. This subsection should describe all required external interface requirements. This should include references to any existing interface control documentation. (Note: this subsection is an overview; details of the specific requirements will be contained in Section 4 of the SRS.)

Constraints. This subsection should provide a list of the general constraints imposed on the system that may limit a designer's choices. Constraints may come from regulations or policy, hardware limitations, interfaces to other applications, parallel operations, audit functions, control functions, higher-order language requirements, communication protocols, criticality of the applications or safety, and security restrictions. This section should not describe the implementation of the constraints, but should provide a source for understanding why certain constraints will be expected from the design.

Assumptions and Dependencies. This subsection should list all assumptions and dependencies that impact the product of the software resulting from the SRS. This subsection should be the source for recognizing the impact of any changes to the assumptions or dependencies on the SRS and resulting software. This section can highlight unresolved requirement issues and should be recorded on the project manager's open issues list.

Requirements Management. This subsection should provide an overview of the requirements management process and any associated procedures. References to existing organizational requirements management plans may be used.

Resources and Funding. This subsection should identify individuals responsible for requirements management. If this information were provided in an existing requirements management plan, a reference to this plan would be provided here.

Reporting Procedures. This subsection should describe all reporting procedures associated with requirements management activities. Details should be provided for the reporting procedures and reporting schedule. If this information is provided in an organizational software requirements management plan, or if this information is provided in the software project management plan, then a reference to these plans should be provided.

Training. This subsection should either reference an organizational training plan, which describes the method and frequency of requirements elicitation, documentation, and management training, or provide the details for all associated training.

Specific Requirements. This section should contain all details that the software developer needs to create a design. The details within this section should be defined as individual, specific requirements with sufficient cross-referencing back to related discussions in the Introduction and Overall Description sections mentioned above. The ordering of subsections shown below is one of several possible combinations that may be selected. The ordering below should normally be used, but others may be selected, provided that they are justified. Each specific requirement should be stated such that its achievement can be objectively verified by a prescribed method. The method of verification will be shown in a requirements traceability matrix in Appendix A of the SRS.

Acceptance. This section should describe the acceptance process, referencing any associated documentation (e.g., software test plan).

Documentation. This section should list any additional supporting process or project documentation not previously referenced.

Maintenance. This section is required if the system must meet requirements for continued operation and enhancement.

Traceability Matrix. This is a comprehensive baseline of all software requirements described within the SRS. This list should be used as the basis for all requirements management, software design, development, and testing. Please refer to Appendix C, Software Process Work Products, for additional information on supporting a requirements traceability matrix.

System Requirements Specification

The information provided here in support of requirements management is designed to facilitate the definition of a system requirements specification. This information was developed using IEEE Std 1233-1998, IEEE Guide for Developing System Requirements Specifications [25] that has been adapted to support CMMI® requirements. The modification of the recommended SysRS table of contents to support the goals of the CMMI®-SW methodology more directly could look like that in Table 6-3.

System Requirements Specification

The following provides section-by-section guidance in support of the creation of a system requirements specification (SysRS). This guidance should be used to help establish a requirements baseline and should reflect the actual processes and procedures of the implementing organization. Additional information is provided in the document template, *System Requirements Specification.doc*, which is located on the CD-ROM accompanying this book.

Introduction. This section should explain the purpose and scope of the project system requirements specification (SysRS), as well as provide clarification of definitions, acronyms, and references. This section should also provide an overview of the project. Subsections are:

System Purpose. This should explain the purpose for writing the SysRS for this project and describe the intended audience for the SysRS. (Note: this maybe aligned with the ConOps document System Overview.)

System Scope. This should identify the system products to be produced, by name; explain what the system products will and will not do; and describe the application of the system being specified, including all relevant goals, objectives, and benefits from producing the system. (Note: this maybe aligned with the ConOps document System Overview.)

Definitions, Acronyms, and Abbreviations. This should provide the definitions of all terms, acronyms, and abbreviations required to properly interpret the SysRS. (Note: this maybe aligned with the ConOps document Definitions and Acronyms.)

Key References. Should list all references used within the SysRS.

System Overview. This should describe what the rest of the SysRS contains as it relates to the systems and software components effort. (Note: this maybe aligned with the ConOps document System Overview.) It should also explain how the document is organized.

General System Description. This should describe the general factors that affect the product and its requirements. (Note: this maybe aligned with the ConOps document Proposed System.) Subsections are:

System Context. This should include diagrams and narrative to provide an overview of the context of the system, defining all significant interfaces crossing the system's boundaries.

System Modes and States. This should include diagrams and narrative to provide usage/operation modes and transition states. (Note: this maybe aligned with the ConOps document Modes of Operation.)

Table 6-3. System requirements specification document outline

Title Page
Revision Page
Table of Contents
1. Introduction
    1.1 System Purpose
    1.2 System Scope
    1.3 Definitions, Acronyms, and Abbreviations
    1.4 References
    1.5 System Overview
2. General System Description
    2.1 System Context
    2.2 System Modes and States
        2.2.1 Configurations
    2.3 Major System Capabilities
    2.4 Major System Conditions
    2.5 Major System Constraints
    2.6 User Characteristics
    2.7 Assumptions and Dependencies
    2.8 Operational Scenarios
3. System Capabilities, Conditions, and Constraints
    3.1 Physical
        3.1.1 Construction
        3.1.2 Durability
        3.1.3 Adaptability
        3.1.4 Environmental Conditions
    3.2 System Performance Characteristics
        3.2.1 Load
        3.2.2 Stress
        3.2.3 Contention
        3.2.4 Availability
    3.3 System Security and Safety
    3.4 Information Technology Management
    3.5 System Operations
        3.5.1 System Human Factors
        3.5.2 System Usability
        3.5.3 Internationalization
        3.5.4 System Maintainability
        3.5.5 System Reliability
    3.6 Policy and Regulation
    3.7 System Life Cycle Sustainment
4. System Interfaces
5. Specific Requirements

Configurations. This should describe typical system configurations that meet various customers' needs.

Major System Capabilities. This should provide diagrams and accompanying narrative to show major capability groupings of the requirements.

Major System Conditions. This should show major conditions, relative to the attributes of the major capability groupings.

Major System Constraints. This should show major constraints, relative to the boundaries of the major capability groupings.

User Characteristics. This should identify each type of user of the system (by function, location, and type of device), the number in each group, and the nature of their use of the system. (Note: this maybe aligned with the ConOps document User Classes.)

Assumptions and Dependencies. This should address all assumptions and dependencies that impact the system resulting from the SysRS. It is important to tie assumptions to assessments of impact. This subsection should be the source for recognizing the impact of any changes to the assumptions or dependencies on the SysRS and resulting system. This section can highlight unresolved requirements issues and should be recorded on the project manager's open issues list.

Operational Scenarios. This should provide descriptive examples of how the system will be used. Scenarios may also be used to describe what the system will not do. Scenarios should be used to help readers understand how system functionality will support operational requirements. (Note: this maybe aligned with the ConOps document Operational Scenarios.)

System Capabilities, Conditions, and Constraints. This should describe the next levels of system structure. System behavior, exception handling, manufacturability, and deployment should be covered under each capability, condition, and constraint. Subsections are:

Physical

Physical Construction. This should include the environmental (mechanical, electrical, chemical) characteristics of the location in which the system will be installed.

Physical Dependability. This should include the degree to which a system component is operable and capable of performing its required function at any (random) time, given its suitability for the mission and whether the system will be available and operate as many times and as long as needed. Examples of dependability measures are availability, interoperability, compatibility, reliability, repeatability, usage rates, vulnerability, survivability, penetrability, durability, mobility, flexibility, and reparability.

Physical Adaptability. This should include how to address growth, expansion, capability, and contraction.

Environmental Conditions. This should include environmental conditions to be encountered by the system. The following subjects should be considered for coverage: natural environment (wind, rain, temperature); induced environment (motion, shock, noise); and electromagnetic signal environment.

System Performance Characteristics. This should include the critical performance conditions and their associated capabilities. Performance requirements for the operational phases and modes should be considered, as should endurance capabilities, minimum total life expectancy, operational session duration, planned utilization rate, and dynamic actions or changes.

Load. This should describe the range of work or transaction loads the system will operate or process in one time period.

Stress. This should describe the extreme but valid conditions for system work or transaction loads.

Contention. This should identify any sharing of computer resources whereby two or more processes can access the computer's processor simultaneously.

Availability. This should describe the range of operational availability or the actual availability of a system under logistics constraints.

System Security and Safety. This should cover both the facility that houses the system and operational security requirements, including requirements of privacy, protection factors, and safety.

Information Technology Management. This should describe the various stages of information processing from production to storage and retrieval to dissemination for the better working of the user organization.

System Operations. This should describe how the system supports the user needs and the needs of the user community.

System Human Factors. This should include references to applicable documents and specify any special or unique requirements for personnel and communications and personnel/equipment interactions.

System Usability. This should describe the special or unique usability requirements for personnel and communications and personnel/equipment interactions.

Internationalization. For each major global customer segment, this should describe the special or unique internationalization requirements for personnel and communications and personnel/equipment interactions.

System Maintainability. This should describe in quantitative terms, the requirements for the planned maintenance, support environment, and continued enhancements.

System Reliability. This should specify in quantitative terms, the reliability conditions to be met. Consider including the reliability apportionment model to support allocation of reliability values assigned to system functions for their share in achieving desired system reliability.

Policy and Regulation. This should describe relevant organizational policies that will affect the operation or performance of the system, including any external regulatory requirements or constraints imposed by normal business practices.

System Life Cycle Sustainment. This should describe quality activities and measurement collection and analysis, to help sustain the system through its expected life cycle.

System Interfaces. This should include diagrams and narrative to describe the interfaces among different system components and their external capabilities, including all its users, both human and other systems. Interfaces to interdependencies or constraints should be included.

Specific Requirements. This section should contain all details that the system/software developer needs to create a design. The details within this section should be defined as individual, specific requirements with sufficient cross-referencing to related discussions in the Introduction and Overall Description sections noted above. The ordering of subsections shown below is one of several possible combinations that may be selected. The ordering below will normally be used, but others may be selected, provided that they are justified. Each specific requirement should be stated such that its achievement can be objectively verified by a prescribed method. Each requirement should enumerated for tracking. The method of verification will be shown in a requirements traceability matrix in Appendix A of the SySRS.

Traceability Matrix. This is a comprehensive baseline of all system requirements described within the SysRS. This list should be used as the basis for all requirements management, system and allocated software requirements, design, development, and testing.

Concept of Operations

The following information is based on IEEE Std 1362-1998, IEEE Guide for Information Technology—System Definition—Concept of Operations (ConOps) Document [28]. The following provides section-by-section guidance in support of the creation of a ConOps document. This guidance should be used to help define a requirements management process and should reflect the actual processes and procedures of the implementing organization. IEEE Std 12207.0 provides information in support of the development of a concept of operations description. The information provided in this section is in conformance with IEEE Std 12207.

Additional information is provided in the document template, *ConOpsDocument.doc,* which is located on the CD-ROM accompanying this book. Table 6-4 provides an example of a suggested document outline supporting CMMI® requirements.

A ConOps document is used to represent the viewpoint of the system, or software, user. The ConOps should effectively describe the system characteristics to all participants and how these characteristics meet the mission and objectives of the organization.

Document Overview. A description of the document, the target audience, and the rationale for its use should be presented in this section.

System Overview. This section of the document should provide an overview of the system and the intended audience and purpose. A diagram of the system, providing an overview of the functionality, is often useful when trying to convey this type of information.

Referenced Documents. This section should provide all supporting documentation used in the development of this document. All documentation referenced by the ConOps should also be listed in this section.

Definitions and Acronyms. This section should provide a list of all definitions and acronyms unique to this document and critical to understanding its content.

Operating Procedures. This section should describe the current operating procedures. These procedures can be based upon an existing system or may describe manual procedures. If there are no current operating procedures, then the ideal should be described. Subsections are:

Background. This section should provide the reader with information in support of the project background, objectives, and scope. The information provided here should describe the problem set and proposed solution.

Operational Policies and Constraints. This section should provide information in support of all operational policies and constraints as they apply to the current system or situation. As defined by IEEE Standard 1362-1998, IEEE Guide for Information Technology—System Definition—Concept of Operations (ConOps) Document, "Policies limit decision-making freedom but do allow for some discretion. Operational constraints are limitations placed on the operations of the current system. Examples of operational constraints include the following:

- A constraint on the hours of operation of the system, perhaps limited by access to secure terminals.
- A constraint on the number of personnel available to operate the system.
- A constraint on the computer hardware (for example, must operate on computer X).
- A constraint on the operational facilities, such as office space."

Table 6-4. Concept of operations (ConOps) document outline [28]

Title Page
Revision Page
Table of Contents
1. Scope
    1.1 Document Overview
    1.2 System Overview
2. Referenced Documents
3. Definitions and Acronyms
4. Operating Procedures
    4.1 Background
    4.2 Operational Policies and Constraints
    4.3 Current Operating Procedures
    4.4 Modes of Operation
    4.5 User Classes
    4.6 Support Environment
5. Change Justification
    5.1 Changes Considered
    5.2 Assumptions and Constraints
6. Proposed System
7. Operational Scenarios
8. Impact Summary
    8.1 Operational Impact
    8.2 Organizational Impact
9. System Analysis
    9.1 Summary of Improvements
    9.2 Disadvantages and Limitations
    9.3 Alternatives
Appendices

Current Operating Procedures. This section should provide a detailed description of the current system or situation. This should include a description of the operating environment, major system components, required interfaces, performance requirements, and desired features. Presenting this information to the system user in graphical format is useful when attempting to describe a system or situation. It may be useful to include items such as schedules, charts, functional and/or data flow, or workflow diagrams. This section should also provide a description of all operational requirements in language that the system user would understand. All required facilities, material, hardware and software, and personnel should be described.

Modes of Operation. This section should provide detailed information describing all required modes of operation for the system or situation.

User Classes. This section should provide a description of all user classes. This should include a description of the organizational structure, user profiles with associated roles and required skill sets, and required interaction between user classes. This section should also describe all key personnel associated with the project. Behavior diagrams can include the use-case diagram (used by some methodologies during requirements gathering), sequence diagram, activity diagram, collaboration diagram, and statechart diagram.

Support Environment. As applicable, this section should describe all support activity required for the maintenance of the system or situation.

Change Justification. This section should provide a description of all deficiencies of the current system or situation. It should provide a description of all identified problems in terms that can be easily understood by the user. This section should also provide a description of all known issues, proposed changes, and justification for the proposed changes. It is helpful to also provide an Appendix to this document that lists all proposed changes in order of importance. A common method used is to categorize items as essential (must have), desired (nice to have), or optional. This ranking is helpful during the translation of the items described in the ConOps document to a more formal requirements specification. Subsections are:

Changes Considered. This section should provide information regarding all features (changes) considered but not included in the proposed software application. The impact of not including these features should be described along with any plans for future adoption.

Assumptions and Constraints. This section should address all assumptions and constraints that will affect users during development and operation of the software application. It is important to tie assumptions to assessments of impact. It should describe what the new system will provide in terms of performance gains, interfaces to external systems, or schedule impact.

Proposed System. This section should provide a high-level, broad system description of the proposed system. This description should be solution oriented. It should focus on the user needs and how the proposed system will support the needs of the user community. Any description of the proposed system should address the characteristics of the operating environment, the performance characteristics, all interface requirements, the capabilities and functions, a description of the data flow requirements, associated cost and risk, and quality requirements.

It is important to keep in mind that the ConOps should be written using common language. Avoid computer related jargon and use graphics where possible. All items described in Section 3 of this document should be addressed in this section.

Operational Scenarios. This section should provide a description, or series of descriptions, of how the proposed system will operate. Scenarios may also be used to describe what the system will not do. Scenarios should be used to help readers understand how system functionality will support operational requirements.

Impact Summary. This section should describe the positive and negative perceived impacts of the proposed system. All predeployment, deployment, and training activity should be described and any impact description provided. Providing this type of information will help organizations prepare for any disruptions caused during system deployment. Subsections are:

> *Operational Impact.* This section should describe all anticipated impacts to the system user during system operation. System deployment may require changes to current policy or procedure, and these should be addressed.
>
> *Organizational Impacts.* This section should describe all anticipated organizational impacts. All impacts associated with users, system development, and system deployment during the operation of the proposed system should be addressed.

System Analysis. This section should provide an analysis of the proposed system. A summary of all improvements, disadvantages and limitations, and alternatives considered as relating to the proposed system should be described. Subsections are:

> *Summary of Improvements.* This section should provide an evaluation of all improvements to existing processes or practices to be provided by deployment of the proposed system. This summary should include a description of any new or enhanced capabilities.
>
> *Disadvantages and Limitations.* This section should describe any perceived disadvantages or limitations presented by the deployment of the proposed system.
>
> *Alternatives.* This section should present a description of any alternatives considered. All alternatives considered, but rejected, should be documented with a rationale for nonacceptance.

Appendices. To facilitate ease of use and maintenance of the ConOps document, some information may be placed in appendices to the document. Each appendix should be referenced in the main body of the document where that information would normally have been provided.

TECHNICAL SOLUTION

The Technical Solution (TS) process area (PA) converts the requirements into the product architecture, product-component design, and the product component itself (e.g., coding).

TS is part of the Engineer Process Category, which contains five other PAs: Requirements Management, Requirements Development, Product Integration, Verification, and Validation.

The TS receives requirements from Requirements Development in order to design, develop, and implement solutions to requirements. Specifically, TS

- Evaluates and selects solutions that potentially satisfy an appropriate set of allocated requirements.
- Develops detailed designs for the selected solutions.
- Implements the designs into code as a product or product component.

TS inputs are from Requirements Development, and Verification, Validation, and Product Integration use the TS outputs. Subsequently, TS receives errors to correct from Verification, Validation, and Product Integration. TS operate under the management of Project Plan, Project Monitoring and Control, and Requirements Management PAs. TS design reviews are covered in Verification. The task of selecting the final solution makes use of the specific practices in the Decision Analysis and Resolution PA. Of all CMMI® PAs, normally the largest numbers of developers are assigned to the TS PA.

The TS PA results in several typical work product or artifacts, some which are expected in other PAs. IEEE Standards 1016, IEEE Recommended Practice for Software Design Descriptions [11]; 1471, IEEE Recommended Practice for Architectural Description of Software Intensive Systems [33]; 1320.1, IEEE Standard for Functional Modeling Language-Syntax and Semantics for IDEF0 [26]; 1320.2, IEEE Standard for Conceptual Modeling Language—Syntax and Semantics for IDEF1X97 [27]; 1420.1a, Data Model for Reuse Library Interoperability: Asset Certification Framework [28]; 1420.1b, Data Model for Reuse Library Interoperability: Intellectual Property Rights Framework [29]; 1008, IEEE Standard for Software Unit Testing [8]; and 1063, IEEE Standard for Software User Documentation [18] provide detailed guidance in support of the development of the designs, unit test plans, and user documents. Additional information is provided in Appendix C describing the work products associated with the TS PA.

IEEE Std 12207.0, Standard for Information Technology—Software life cycle processes [39], describes 17 processes spanning the entire life cycle of a software product or service. Even if an organization's processes were defined using other sources, the standard is useful in characterizing the essential characteristics of these software processes and should be considered prior to the implementation of process improvement activities. Referencing this standard and reviewing what is required for each of these primary process areas can provide additional guidance in support of the activities associated with the requirements as defined by the technical solution PA. It is important that IEEE 12207 be considered prior to the implementation of any process improvement activities associated with technical solution.

CMMI®-SW Goals

The CMMI® Special Terminology for TS includes Development and Product Component. TS has three specific goals and one general goal:

SG 1. Product or product-component solutions are selected from alternative solutions.
    Alternative solutions are considered when selecting a solution. Reuse and use of

commercial off-the-shelf (COTS) product components are also considered. Architectural features are important as they provide a foundation for product improvement and evolution. This goal involves Requirements Development and usually results in a revised requirement document.

*SG 2. Product or product-component designs are developed.*Product or product-component designs provide appropriate direction for code implementation, and also for other activities of the product life cycle such as procurement, maintenance, sustainment, and installation.

SG 3. Product components, and associated support documentation, are implemented from their designs. Product components are code implemented from the designs. The implementation includes unit testing of the product components before sending them to product integration and development of end-user documentation.

GG 3. The process is institutionalized as a defined process. CMMI®-SW specifically requires the TS PA be institutionalized as a defined process, so the project team can tailored their processes and document templates from the organization's set of standard processes according to the organization's tailoring guidelines. Table 6-5 describes the specific and generic goals and practices, typical work products, and commonly associated plans or artifacts in support of the PA.

Design Document

The following information is primarily based on IEEE Std 1016-1998, IEEE Recommended Practice for Software Design Descriptions [11] and IEEE 12207.1-1997 [39], Standard for Information Technology—Software Life Cycle Processes—Life Cycle Data. IEEE Std 1471-2000, IEEE Recommended Practice for Architectural Description of Software Intensive Systems [33]; IEEE Std 1320.1-1998, IEEE Standard for Functional Modeling Language—Syntax and Semantics for IDEF0 [26]; and IEEE Stdards 1420.1-1995, 1420.1a-1996, and 1420.1b-2002, Software Reuse—Data Model for Reuse Library Interoperability [29] were also used as source documents for the development of this information. All standards used have been adapted to support CMMI requirements. Table 6-6 provides a sample document outline supporting CMMI® requirements.

The following provides section-by-section guidance in support of the creation of an SDD. This guidance should be used to help define associated design decisions and should reflect the actual requirements of the implementing organization. Additional information is provided in the document template, *Software Design Document.doc,* which is located on the CD-ROM accompanying this book.

Introduction. As stated in IEEE 1016, "The SDD shows how the software system will be structured to satisfy the requirements identified in the software requirements specification. It is a translation of requirements into a description of the software structure, software components, interfaces, and data necessary for implementation. In essence, the SDD becomes a detailed blueprint for the implementation activity."

This section should explain the purpose and scope of project software design, as well as provide clarification of definitions, acronyms, and references. This section should also provide an overview of the project. Subsections are:

Purpose. This subsection should explain the purpose for writing an SDD for this project and describe the intended audience for the SDD.

Table 6-5. Technical solution goals and practices [61]

Specific and generic goals/practices and typical work products	Plan or artifact
SG1. Select Product-Component Solutions	
SP1.1. Develop detailed alternative solutions and selection criteria	
Alternative solution screening criteria	Design documentation, alternative solution screening criteria matrix
Evaluations of new technologies	Design documentation
Alternative solutions	Design documentation
Selection criteria for final selection	Design documentation
SP1.1-1. Develop alternative solutions and selection criteria (continuous)	
Alternative solutions	Design documentation
Selection criteria	Design documentation
SP 1.2. Evolve operational concepts and scenarios	
Product-component operational concepts, scenarios, and environments for all product-related life-cycle processes	Concept of operations document
Timeline analyses of product-component interactions	Concept of operations document
Use cases	Software requirements specifications, UML modeling
SP 1.3. Select product-component solutions	
Product-component selection decisions and rationale	Design documentation
Documented relationships between requirements and product components	Requirements traceability matrix
Documented solutions, evaluations, and rationale	Design documentation
SG2. Develop the Design	
SP 2.1. Design the product or product component	
Product architecture	Design documentation, Architecture design (IEEE 1471)
Product-component designs	Design documentation
SP 2.2. Establish and maintain a technical data package	
Technical data package	Software configuration management plan
SP 2.3. Design interfaces using criteria	
Interface design specifications	Interface control document
Interface control documents	Interface control document
Interface specification criteria	Interface control document
Rationale for selected interface design	Interface control document
SP 2.3-1. Establish interface descriptions (continuous)	
Interface design	Interface control document
Interface design documents	Interface control document
SP 2.4. Perform make, buy, or reuse analyses	
Criteria for design and product-component reuse	Make/buy decision matrix
Make-or-buy analyses	Make/buy decision matrix
Guidelines for choosing COTS product components	Make/buy decision matrix

(continued)

Table 6-5. *Continued*

Specific and generic goals/practices and typical work products	Plan or artifact
SG3. Implement the Product Design	
SP 3.1. Implement the design	
Implemented design	Software code, data structures, peer inspection report descriptions, program code walk-through checklist, unit test report,
SCM records	
SP 3.2. Develop product support documentation	Postdevelopment WBS
End-user training materials	Associated manual and/or training material
User's manual	User's manual
Operator's manual	Associated manual and/or training material
Maintenance manual	Associated manual and/or training material
Online help	Associated manual and/or training material
GG3. Institutionalize a Defined Process	
GP 2.1. Establish an organizational policy	
Establishes organizational expectations for addressing the iterative cycle in which product-component solutions are selected, product and product-component designs are developed, and the product-component designs are implemented	Technical solution policy
GP 2.2. Plan the process	Software project management plan
GP 2.3. Provide resources	
Identification of resources used in support of implementation activities	Software project management plan
GP 2.4. Assign responsibility	Software project management plan, requirements management plan
GP 2.5. Train people	Software project management plan or training plan
GP 2.6. Manage configurations	
Description of how work products and all associated artifacts in support of software development activities are placed under configuration management	Configuration management plan
GP 2.7. Identify relevant stakeholders	
Describe activities and feedback mechanisms	Software project management plan
GP 2.8. Monitor and control the process	Software project management plan or software measurement and metrics plans
GP 2.9. Objectively evaluate adherence	Quality assurance plan
GP 2.10. Review status with higher-level management	Software project management plan
GP 3.1. Establish a defined process	
Establish and maintain a description of the technical solution process	Software project management plan
GP 3.2. Collect Improvement Information	All above, typical work products

Table 6-6. Software design document outline

Title Page
Revision Page
Table of Contents
1. Introduction
    1.1 Purpose
    1.2 Scope
    1.3 Definitions, Acronyms, and Abbreviations
    1.4 References
2. Design Overview
    2.1 Background Information
    2.2 Alternatives
3. User Characteristics
4. Requirements and Constraints
    4.1 Performance Requirements
    4.2 Security Requirements
    4.3 Design Constraints
5. System Architecture
6. Detailed Design
    6.1 Description for Component N
    6.2 Component N Interface Description
7. Data Architecture
    7.1 Data Analysis
    7.2 Output Specifications
    7.3 Logical Database Model
    7.4 Data Conversion
8. Interface Requirements
    8.1 Required Interfaces
    8.2 External System Dependencies
9. User Interface
    9.1 Module [X] Interface Design
    9.2 Functionality
10. Nonfunctional Requirements
Traceability Matrix

Scope. This subsection should:

- Identify the software products to be produced, by name
- Explain what the software products will, and if necessary, will not do
- Describe the application of the software being specified, including all relevant goals, objectives, and benefits from producing the software.

Provide a description of the dominant design methodology. Provide a brief overview of the product architecture. Briefly describe the external systems with which this system must interface. Also explain how this document might evolve throughout the project lifecycle.

Definitions, Acronyms, and Abbreviations. This subsection should provide the definitions of all terms, acronyms, and abbreviations required to properly interpret the SDD.

References. This subsection should list all references used within the SDD. All relationships to other plans and policies should be described as well as existing design standards.

Design Overview. Subsections are:

> *Background Information.* This section should briefly present background information relevant to the development of the system design. All stakeholders should be identified and their contact information provided. A description of associated project risks and issues may also be presented along with assumptions and dependencies critical to project success. Describe the business processes that will be modeled by the system.

> *Alternatives.* All design alternatives should be considered and the rationale for nonacceptance should be briefly addressed in this section. See Appendix C, Make/Buy/Mine/Commission Decision Matrix and Alternative Solution Screening Criteria Matrix sections.

User Characteristics. This section should identify the potential system users, specify the levels of expertise needed by the various users and indicate how each user will interact with the system, and describe how the system design will meet specific user requirements.

Requirements and Constraints. Subsections are:

> *Performance Requirements.* This section should describe how the proposed design will ensure that all associated performance requirements will be met.

> *Security Requirements.* This section should describe how the proposed design will ensure that all associated security requirements will be met, list any access restrictions for the various types of system users, describe any access code systems used in the software, identify any safeguards that protect the system and its data, and specify communications security requirements.

> *Design Constraints.* This section should describe how requirements that place constraints on the system design would be addressed and list all dependencies and limitations that may affect the software. Examples include budget and schedule constraints, staffing issues, availability of components, and so on.

System Architecture. This section should provide a description of the architectural design. See Appendix C, Architecture Design Success Factors and Pitfalls. All entities should be described as well as their interdependent relationships. A top-level diagram may be provided. See Appendix C, Unified Modeling Language (UML), for an example showing inheritance, aggregation, and reference relationships.

Detailed Design. Break the system down into design components that will interact with and transform data to perform the system objectives. Assign a unique name to each component, and group these components by type, for example, class, object, or procedure. Describe how each component satisfies system requirements. In user terminology, specify the inputs, outputs, and transformation rules for each component. Depict how the components depend on one another. Each component should be described as shown by the following list of suggested headers:

- Description for Component N
- Processing Narrative for Component N

- Component N Interface Description
- Component N Processing Detail

Data Architecture. This section should describe the data structures to be used in support of the implementation. If these include databases, define the table structure, including full field descriptions, relationships, and critical database objects. Graphical languages are appropriate. This information is often provided in a separate database design document. If this is the case simply refer to this document and omit the remainder of this section. Subsections are:

> *Data Analysis.* A brief description of the procedures used in support of data analysis activities should be described in this section. Any analysis of the data that resulted in a change to the system design, or that impacted system design, should be noted.

> *Output Specifications.* All designs supporting requirements for system outputs should be described in this section. These may include designs to support reporting, printing, e-mail, and so on.

> *Logical Database Model.* Identify specific data elements and logical data groupings that are stored and processed by the design components in Section 6. Outline data dependencies, relationships, and integrity rules in a data dictionary. Specify the format and attributes of all data elements or data groupings. A logical model of data flow, depicting how design elements transform input data into outputs, should be developed and presented here.

> *Data Conversion.* This section should describe all design requirements in support of data conversion activities. This may be covered in a separate data migration plan, but if it is not, it should be documented in the design documentation. The migration and validation of any converted legacy data should be described.

Interface Requirements. Subsections are:

> *Required Interfaces.* This section should describe all interfaces required in support of hardware and software communications. If an interface control document was developed along with the SRS, it should be referenced here. The effectiveness of how the system design addresses relevant interface issues should be discussed. Specify how the product will interface with other systems. For each interface, describe the inputs and outputs for the interacting systems. Explain how data is formatted for transmission and validated upon arrival. Note the frequency of data exchange.

> *External System Dependencies.* This section should provide a description of all external system dependencies. A diagram may be used to provide a description of the system with each processor and device indicated.

User Interface. This section should describe the user interface and the operating environment, including the menu hierarchy, data entry screens, display screens, online help, and system messages. It should specify where in this environment the necessary inputs are made, and list the methods of data outputs, for example, printer, screen, or file. Subsections are:

[Module X] Interface Design. This section should contain screen images and a description of all associated design rules.

Functionality. All objects and actions should be described in this section. Any reuse of existing components should be identified.

Nonfunctional Requirements. This section should address design items associated with nonfunctional requirements relating to system performance, security, licensing, language, or other related items.

Traceability Matrix. It is necessary to show traceability throughout the product lifecycle. All design items should be traceable to original system requirements. This can be annotated in the requirements traceability matrix, the design review document, or in the configuration management system.

User's Manual

The following provides section-by-section guidance in support of the creation of a software user's manual or operator's manual. IEEE Std 1063-2001, IEEE Standards for Software User Documentation [18] was the primary reference document for the development of this material.

Users Manual Document Guidance

This guidance should be used to help define a management process and should reflect the actual processes and procedures of the implementing organization. Additional information is provided in the document template, *Software Users Manual.doc,* which is located on the CD-ROM accompanying this book. Table 6-7 provides an example document outline.

Introduction. This section should provide information on document use, all definitions and acronyms, and references. Subsections are:

   Document Use. This section should describe the intended use of the software user's manual. The organization of the user's manual should effectively support its use. If the user's manual is going to contain both instructional and reference material, each type should be clearly separated into different chapters or topics. Task-oriented documentation (instructional) should include procedures that are structured according to user's tasks. Documentation used as reference material should be arranged to provide access to individual units of information. This section can provide an overview of the type of information provided, its intended use, and the organization of the user's manual.

   Definitions and Acronyms. This section should identify all definitions and acronyms specific to this software user's manual. This should be an alphabetical list of application-specific terminology. All terminology used with the users manual should be consistently applied.

   References. This section should provide a list of all references used in support of the development of the software user's manual. Include a listing of all the documentation related to the product that is to be transitioned to the operations area, which in-

Table 6-7. User's manual document outline

Title Page
Revision Page
Table of Contents
1. Introduction
1.1 Document Use
1.2 Definitions and Acronyms
1.3 References
2. Concept of Operations
3. General Use
4. Procedures and Tutorials
5. Software Commands
6. Navigational Features
7. Error Messages and Problem Resolution
Index

cludes any security or privacy protection consideration associated with its use. Also include as a part this any licensing information for the product.

Concept of Operations. This section should provide an overview of the software, including its intended use. Descriptions of any relevant business processes or workflow activities should be included. Any items required in support of the understanding of the software product should be included. This may require a description of theory, method, or algorithm critical to the effective use and understanding of the product.

General Use. Information should be provided in support of routine user activities. It is important to identify actions that will be performed repetitively to avoid redundancy within the user's manual. For example, describing how to cancel or interrupt an operation while using the software would be in this section. Other task-oriented routine documentation could include software installation and deinstallation procedures, how to log on and off the application, and the identification of basic items/actions that are common across the applications' user interface.

Procedures and Tutorials. Information of a tutorial (i.e., procedural) nature should be provided in the user's manual as clearly as possible. A consistent approach to the presentation of the material is important when trying to clearly communicate a concept to the user.

Describe the purpose and concept for the tutorial information presented in the user's manual. Include a list of all activities that must be completed prior to the initiation of the procedure or tutorial. Identify any material that should be used as reference in support of the task. List all cautions and other supporting information that are relevant in supporting the performance of the task.

It is important to list all instructional steps in the order in which they should be performed, with any optional steps clearly identified. The steps should be consecutively numbered and the initial and last steps of the task should be clearly identified. It is important that the user understands how to successfully initiate and complete the procedure or tutorial.

Warnings and cautions should be distinguishable from instructional steps and should be preceded by a word and graphic symbol alerting the user to the item. For example, *warning (graphic),* would precede a warning to the user. The use of the following format for warning and cautions is suggested: word and graphic, brief description, instructional text, description of consequences, and proposed solution or workaround.

Software Commands. The user's manual should describe all software commands, including required and optional parameters, defaults, precedence, and syntax. All reserved words and commands should be listed. This section should not only provide the commands, but should also provide examples of their use. Documentation should include a visual representation of the element, a description of its purpose, and an explanation of intended action. A quick reference card may be included in the user's manual, providing the user with the ability to rapidly refer to commonly used commands.

Navigational Features. The document should describe all methods of navigation related to the software application. All function keys, graphical user interface items, and commands used in support of application navigation should be described and supported with examples.

Error Messages and Problem Resolution. Information in support of problem resolution (i.e., references) should address all known problems or error codes present in the software application. Users should be provided information that will either help them recover from known problems, report unknown issues, or suggest application enhancements.

Index. An index provides an effective way for users to access documented information. It is important to remember that for an index to be useful it should contain words that users are most likely to look up and should list all topics in the user's manual. Pay special attention to the granularity and presentation of the index topics. Place minor key words under major ones; for example, instead of using *files* with 30 pages listed, use *files, saving* and *files, deleting* with their associated specific pages listed.

Software Transition Plan

The purpose of transition planning is to lay out the tasks and activities that need to take place to efficiently move a product (i.e., specify product name, or in-house developed software, or COTS software, middleware, or component software/hardware) from the development or pilot environment to the production, operations, and maintenance environment. The transition planning steps apply whether the product is being transitioned within an agency (i.e., from agency development staff to agency network or operations staff) or to an outside agency (i.e., information technology systems).

The transition plan is designed to facilitate migration of an application system from development to production (i.e., maintenance). The information provided here in support of transition plan development is based upon IEEE Std-1219, IEEE Standard for Software Maintenance [22], IEEE Std 12207.0, Standard for Information Technology—Life Cycle Processes [39], and U.S. DOD Data Item Description DI-IPSC-81429, Software Transition Plan [76] as primary reference material. Additional information has been incorporated as "lessons learned" from multiple production application systems and transition opportunities. Table 6-8 provides a proposed document outline in support of CMMI® requirements.

Table 6-8. Transition plan document outline

Title Page
Revision Page
Table of Contents
  1. Introduction
      1.1 Overview
      1.2 Scope
      1.3 Definitions and Acronyms
          1.3.1 Key acronyms
          1.3.2 Key terms
      1.4 References
  2. Product
      2.1 Relationships
  3. Strategies
      3.1 Identify Strategy
      3.2 Select Strategy
  4. Transition Schedules, Tasks, and Activities
      4.1 Installation
      4.2 Operations and Support
      4.3 Conversion
      4.4 Maintenance
  5. Resource Requirements
      5.1 Software Resources
      5.2 Hardware Resources
      5.3 Facilities
      5.4 Personnel
      5.5 Other Resources
  6. Acceptance Criteria
  7. Management Controls
  8. Reporting Procedures
  9. Risks and Contingencies
  10. Transition Team Information
  11. Transition Impact Statement
  12. Plan Review Process
  13. Configuration Control

Transition Plan Document Guidance

The following provides section-by-section guidance in support of the creation of a software transition plan. The development of a software transition plan is critical to the successful transition of software from development to deployment. This guidance should be used to help develop and define the transition process and should reflect the actual processes and procedures of the implementing organization. Additional information is provided in the document template, *Software Transition Plan.doc,* which is located on the CD-ROM accompanying this book.

Overview. The transition plan should include an introduction addressing background information on the project. It should show the relationship of the project to other projects and/or organizations or agencies, address maintenance resources required, and identify the transition team's organization and responsibilities, as well as the tools, techniques, and methodologies that are needed to perform an efficient and effective transition.

The transition plan should include deployment schedules and resource estimates, and identify special resources and staffing. The transition plan shall also define management controls and reporting procedures, as well as the risks and contingencies. Special attention must be given to minimizing operational risks. An impact statement should be produced outlining the potential impact of the transition to the existing infrastructure, operations, and support staff and to the user community.

Scope. This section should include a statement of the scope of the transition plan. Include a full identification of the product to which this document applies, including (as applicable) identification numbers, titles, abbreviations, version numbers, and release numbers. Include product overview, overview of the supporting documentation, and description of the relationship of the product to other related projects and agencies.

Definitions and Acronyms. This section should identify all definitions and acronyms specific to this software configuration management plan.

References. This section should provide a list of all references used in support of the development of the software configuration management plan. Include a listing of all the documentation related to the product that is to be transitioned to the operations area, including any security or privacy protection consideration associated with its use. Include any licensing information for the product.

Product. This section should include a brief statement of the purpose of the product to which this document applies. It should also describe the general nature of the product; summarize the history of development, operation, and maintenance; identify the project sponsor, acquirer, user, developer, vendor, and maintenance organizations; identify current and planned operating sites; and list other relevant documents.

Relationships. This section should describe the relationship(s) of the product being transitioned to any other projects and agencies. The inclusion of a diagram or flow chart to help indicate these relationships is often helpful.

Identify Strategies. This section should identify the transition strategies and tools to be used as part of the transition plan. Identify all the options for moving the product from its present state into production/operations. These options could include:

- Incremental implementation or phased approach
- Parallel execution
- One-time conversion and switchover
- Any combinations of the above

Each option should also identify the advantages and disadvantages, risks, estimated time frames, and estimated resources.

Select Strategy. This section should include an evaluation of each of the transition options, comparing them to the transition requirements, and selecting the one that is most appropriate for the project. Once a transition strategy has been selected, then the justification is documented and approved.

Transition Schedules, Tasks and Activities. This section should include, or provide reference to, detailed schedules for the selected transition strategy. These schedules should include equipment installation, training, conversion, deployment, and/or retirement of the existing system (if applicable) as well as any transition activities required to turn over the product from developers or vendors to operational staff. The schedules should reflect all milestones for conducting transition activities.

Also include an installation schedule for equipment (new or existing), software, databases, and so on. Include provisions for training personnel with the operational software and target computer(s), as well as any maintenance software and/or host system(s). Describe the developer's plans for transitioning the deliverable product to the maintenance organization. This should address the following:

- Planning/coordination of meetings
- Preparation of items to be delivered to the maintenance organization
- Packaging, shipment, installation, and checkout of the product maintenance environment
- Packaging, shipment, installation, and checkout of the operational software
- Training of maintenance/operational personnel

Installation. Installation consists of the transportation and installation of the product from the development environment to the target environment(s). This section should describe any required modifications to the product, checkout in the target environment(s), and customer acceptance procedures. If problems arise, these should be identified and reported; these procedures should be addressed here as well. If known, any temporary "workaround(s)" should also be described.

Operations and Support. This section should address user operations and all required ongoing support activity. Support includes providing technical assistance, consulting with the user, and recording user support requests by maintaining a Support Request Log. The operations and support activities can trigger maintenance activities via the ongoing project monitoring and controlling activities or problem and change logs, and this process should be described or a reference to associated documentation should be provided.

Conversion. This section should address any data or database transfers to the product and its underlying components that would occur during the transition.

Maintenance. Maintenance activities are concerned with the identification of enhancements and the resolution of product errors, faults, and failures. The requirements for software maintenance initiate "service-level changes" or "product modification requests" by using defined problem and change-management-reporting procedures. This section should describe all issues and activities associated with product maintenance during the transition.

Resource Requirements. All estimates for resources (hardware, software, and facilities) as well as any special resources (i.e., service and maintenance contracts) and staffing for the selected transition strategy should be described in this section. The assignment of staff, agency, and vendor responsibility for each task identified should be documented. This allows managers and project team members to plan and coordinate the work of this project with other assignments. If specific individuals cannot be identified when the transition plan is developed, generic names may be used and replaced with individual names as soon as the resources are identified.

Software Resources. A description of any software and associated documentation needed to maintain the deliverable product should be included in this section. The description should include specific names, identification numbers, version numbers, release numbers, and configurations as applicable. References to user/operator manuals or instructions for each item should be included. Identify where each product item is to come from—acquirer furnished, currently owned by the organization, or to be purchased. Include information about vendor support, licensing, and usage and ownership rights, whether the item is currently supported by the vendor, whether it is expected to be supported at the time of delivery, whether licenses will be assigned to the maintenance organization, and the terms of such licenses. Include any required service and maintenance contract costs as well as payment responsibility.

Hardware Resources. This section should include a description of all hardware and associated documentation needed to maintain the deliverable product. This hardware may include computers, peripheral equipment, simulators, emulators, diagnostic equipment, and noncomputer equipment. The description should include specific models, versions, and configurations. References to user/operator manuals or instructions for each item should be included. Identification each hardware item and document as acquirer furnished, items that will be delivered to the maintenance organization, or items the maintenance organization currently owns or needs to acquire. (If the item is to be acquired, include information about a current source of supply, order information, as well as what budget is to pay for it.) Include information about manufacturer support, licensing, usage and ownership rights, whether the items are currently supported by the manufacturer or will be in the future, and whether licenses will be assigned to the maintenance organization and the terms of such licenses.

Facilities. Describe any facilities needed to maintain the deliverable product in this section. These facilities may include special buildings, rooms, mock-ups, building features such as raised flooring or cabling, building features to support security and privacy protection requirements, building features to support safety requirements, special power requirements, and so on. Include any diagrams that may be applicable.

Personnel. This section should include a description of all personnel needed to maintain the deliverable product, include anticipated number of personnel, types of support personnel (job descriptions), skill levels and expertise requirements, and security clearance.

Other Resources. Identify any other consumables (i.e., technology, supplies, and materials) required to support the product. Provide the names, identification numbers, version numbers, and release numbers. Identify if the document or consumable is acquirer furnished, an item that will be delivered to the maintenance organization, an item the organization current owns, or an item the organization needs to acquire. If the maintenance/operational organization needs to acquire it, identify the budget will cover the expense.

Acceptance Criteria. It is important to establish the exit or acceptance criteria for transitioning the product. These criteria will determine the acceptability of the deliverable work products and should be specified in this section. Representatives of the transitioning organization and the acquiring organization should sign a formal agreement, such as a service level agreement, that outlines the acceptance criteria. Any technical processes methods, or tools as well as performance benchmarks required for product acceptance, should be specified in the agreement. Also include an estimation of the operational budget for the product and how these expenses will be covered.

Management Controls. Describe all management controls to ensure that each task is successfully executed and completed based on the approved acceptance criteria. This should include procedures for progress control, quality control, change control, version control, and issue management during the transition process.

Reporting Procedures. This section should define the reporting procedures for the transition period. Include such things as type of evaluations (review, audit, or test) as well as anomalies identified during the performance of these evaluations.

Risks and Contingencies. This section should identify all known risks and contingencies that arise during the transition process, with special attention given to minimizing operational risks. Reference to an organization risk management plan should be provided here. A description of risk mitigation, tracking, and reporting should be presented.

Transition Team Information. This section should include all transition team information, including the transition team's organization, roles and responsibilities for each activity, as well as the tools, techniques, and methodologies and/or procedures that are needed to perform the transition.

Transition Impact Statement. This section should contain a statement that would describe any anticipated impact to existing network infrastructure, support staff, and user community during the system transition. The impact statement should include descriptions for the performance requirements, availability, security requirements, expected response times, system backups, expected transaction rates, initial storage requirements with expected growth rate, as well as help-desk support requirements.

Plan Review Process. This section should describe the review procedures/processes in support of this document. A review should be held to identify and remove any defects from the transition plan before it is distributed. This is a content review and should be conducted by the appropriate members of the project team or an independent third party. The results should be recorded in inspection report and inspection log defect summary, shown in Appendix C.

Configuration Control. The transition plan information should be subject to the configuration control process for the project. Subsequent changes are tracked to ensure that the configuration of the transition plan information is known at all times. Changes shall be allowed only with the approval of the responsible authority. Transition plan changes should follow the same criteria established for the project's change control procedures and reference to any associated project-level configuration management plan should be provided here.

PRODUCT INTEGRATION

The purpose of Product Integration (PI) is to assemble the product from the product components; ensure that the product, as integrated, functions properly; and deliver the product. PI includes the management of products and product components' internal and external interfaces to ensure compatibility among the interfaces. PI is part of the Engineer Process Category, which contains five other PAs: Requirements Management, Requirements Development, Technical Solution, Verification, and Validation.

Requirements Development and Technical Solution determine how the product components are combined and interfaces information. PI uses both Verification and Validation specific practices during product integration implementation. Prior to product integration, Verification verifies the interfaces and interface requirements between product components. Verification specific practices are used during product integration in the operational environment. PI receives tested modules from Verification and integrates them together into a larger subsystem. This can be a recursive approach until the entire system is constructed using Validation specific practices.

Immature organizations use a "big bang" approach for PI, whereas more mature organizations ensure that the product architecture and project plan are tuned to allow early working functionality for early customer feedback.

The PI PA results in several typical work product or artifacts, some which are expected in other PAs. IEEE Std 829, IEEE Standard for Software Test Documentation[5]; IEEE Std 1012, Standard for Software Verification and Validation[9]; IEEE Std 1028, IEEE Standard for Software Reviews [12]; 1320.1, IEEE Standard for Functional Modeling Language [26]; 1044, IEEE Standard Classification for Software Anomalies [13]; and 1465, Adoption of ISO/IEC 12119 Software Packages—Quality Requirements and Testing [32] provide detailed guidance in support of the development of the planning and test results.

IEEE Std 12207.0, Standard for Information Technology—Software Life Cycle Processes [39], describes 17 processes spanning the entire life cycle of a software product or service. Even if an organization's processes were defined using other sources, the standard is useful in characterizing the essential characteristics of these software processes and should be considered prior to the implementation of process improvement activities. Referencing this standard and reviewing what is required for each of these primary process areas can provide additional guidance in support of the activities associated with product integration. It is important that IEEE 12207 be considered prior to the implementation of any process improvement activities associated with product integration.

CMMI®-SW Goals

The goal of PI is to achieve complete integration through the progressive assembly of product components, in one stage or in incremental stages, according to a defined integration sequence and procedures. Attention to internal and external interfaces of the products and product components is critical.

SG 1. Preparation for product integration is conducted.

This goal calls for preparation of the integration sequence, the environment for performing the integration, and integration procedures. Most of this information is reusable on subsequent versions and is developed concurrently with the TS PA.

SG 2. The product-component interfaces, both internal and external, are made compatible.

Product-component interface requirements, specifications, and designs are managed to ensure that implemented interfaces will be complete and compatible.

SG 3. Verified product components are assembled and the integrated, verified, and validated product is delivered.

Product components interfaces are verified and connected according to the defined integration sequence and procedures. Interoperation is also checked. Problems are recorded and their solutions are managed under SCM. Formerly, these practices were called system integration testing.

GG 3. The process is institutionalized as a defined process.

CMMI®-SW specifically requires that the PI PA be institutionalized as a defined process, so the project team can tailor their PI processes and document templates from the organization's set of standard processes according to the organization's tailoring guidelines. Table 6-9 describes the specific and generic goals and practices, typical work products, and commonly associated plans or artifacts in support of the PA.

Interface Control Document

The following information is based on IEEE Std 830-1998, IEEE Recommended Practice for Software Requirements Specifications [6] and The U.S. Dept. of Justice Systems Development Life Cycle Guidance Document, Interface Control Document (ICD) Template* which have been adapted to support CMMI requirements. Additional information is provided in the document template, *Interface Control Document.doc,* which is located on the CD-ROM accompanying this book. Table 6-10 provides an example document outline. Additional information in support of associated PI work products is provided in Appendix C.

The following provides section-by-section guidance in support of the creation of an ICD. This guidance should be used to help define associated interface control requirements and should reflect the actual requirements of the implementing organization.

---

*Department of Justice Systems Development Life Cycle Guidance, Interface Control Document Template, Appendix C-17; http://www.usdoj.gov/jmd/irm/lifecycle/table.htm.

Table 6-9. Product integration goals and practices [61]

Specific and generic goals/practices and typical work products	Plan or artifact
SG1. Prepare for product integration	
SP 1.1. Determine integration sequence	
Product integration sequence	System integration test plan
Rationale for selecting or rejecting integration sequence	System integration test plan
SP 1.2. Establish product integration environment	
Verified environment for product integration	System integration test plan
Support documentation for the product integration environment	System integration test plan
SP 1.3. Establish product integration procedures and criteria	
Product integration procedures	System integration test plan
Product integration criteria	
SG2. Ensure Interface Compatibility	
SP 2.1. Review interface descriptions for completeness	
Categories of interfaces	System integration test plan or interface control documentation
List of interfaces per category	System integration test plan or interface control documentation
Mapping of the interfaces to the product components and product integration environment	System integration test plan or interface control documentation
SP 2.2. Manage interfaces	
Table of relationships among the product components and the external environment	System integration test plan, interface control documentation, or requirements traceability matrix
Table of relationships between the different product components	System integration test plan, interface control documentation, or requirements traceability matrix
List of agreed-to interfaces defined for each pair of product components, when applicable	System integration test plan or interface control documentation
Reports from the interface control working group meetings	System integration test plan
Action items for updating interfaces	System integration test plan
Action items for updating interfaces `	System integration test plan or interface control documentation
Application program interface (API)	System integration test plan or interface control documentation
Updated interface description or agreement	System integration test plan or interface control documentation
SG3. Assemble Product Components and Deliver the Product	
SP 3.1. Confirm Readiness of Product Components for Integration	
Acceptance documents	Unit test report

Table 6-9. Product integration goals and practices [61]

Specific and generic goals/practices and typical work products	Plan or artifact
SG3. Assemble Product Components and Deliver the Product (*cont.*)	
SP 3.1. Confirm Readiness of Product Components for Integration (cont.)	
Delivery receipts	Unit test report
Checked packing lists	Unit test report
Exception reports	Unit test report
Waivers	Unit test report
SP 3.2. Assemble product components	
Assembled product or product components	Integrated software code
	Database structures
SP 3.3. Evaluate assembled product components	
Exception reports	System integration test report, test incident report
Interface evaluation reports	System integration test report or interface control documentation
Product integration summary reports	System integration test report, test summary report
SP 3.4. Package and Deliver the Product or Product Component	
Packaged product or product components	System integration test plan or software configuration management report
Delivery documentation	System integration test plan, software configuration management report, product packaging information
GG3. Institutionalize a Defined Process	
GP 2.1. Establish an organizational policy	
Establishes organizational expectations for developing product integration sequences, procedures, and an environment, ensuring interface compatibility among product components, assembling the product components, and delivering the product and product components	Organizational policy in support of product integration
GP 2.2. Plan the process	System integration test plan
GP 2.3. Provide resources	
Identification of resources used in support of integration activities	System integration test plan
GP 2.4. Assign responsibility	System integration test plan or software project management plan
GP 2.5. Train people	System integration test plan, software project management plan or training plan
GP 2.6. Manage configurations	
Description of how work products and all associated artifacts in support of software integration activities are placed under configuration management	Software configuration management plan

(*continued*)

Table 6-9. Product integration goals and practices [61]

Specific and generic goals/practices and typical work products	Plan or artifact
GG3. Institutionalize a Defined Process (*cont.*) *GP 2.7. Identify relevant stakeholders* Describe activities and feedback mechanisms	System integration test plan or project plan
GP 2.8. Monitor and control the process	Software project management/integration or software measurement and metrics plans
GP 2.9. Objectively evaluate adherence	Reviews are identified in SQA plan, SCM audit
GP 2.10. Review status with higher-level management	Software project management plan
GP 3.1. Establish a defined process	System integration test plan

System Identification. This section should contain a full identification of the participating systems, the developing organizations, responsible points of contact, and the interfaces to which this document applies, including, as applicable, identification numbers(s), title(s), abbreviation(s), version number(s), release number(s), or any version descriptors used. A separate paragraph should be included for each system that is part of the interface.

Table 6-10. Interface control document outline

Title Page
Revision Page
Table of Contents
1. Introduction
    1.1 System Identification
    1.2 Document Overview
    1.3 References
    1.4 Definitions and Acronyms
2. Description
    2.1 System Overview
    2.2 Interface Overview
    2.3 Functional Allocation
    2.4 Data Transfer
    2.5 Transactions
    2.6 Security and Safety
3. Detailed Interface Requirements
    3.1 Interface 1 Requirements
        3.1.1 Interface Processing Time Requirements
        3.1.2 Message (or File) Requirements
        3.1.3 Communication Methods
        3.1.4 Security Requirements
    3.2 Interface 2 Requirements
4. Qualification Methods

Document Overview. This section should provide an overview of the document, including a description of all sections.

References. This section should list the number, title, revision, and date of all documents referenced or used in the preparation of this document.

Definitions and Acronyms. This section should describe all terms and abbreviations used in support of the development of this document and critical to the comprehension of its content.

Description. A description of the interfaces between the associated systems should be described in the following subsections.

> *System Overview.* This subsection should describe each interface and the data exchanged between the interfaces. Each system should be briefly summarized, with special emphasis on functionality relating to the interface. The hardware and software components of each system should be identified.
>
> *Interface Overview.* This subsection should describe the functionality and architecture of the interfacing system(s) as they relate to the proposed interface. Briefly summarize each system, placing special emphasis on functionality. Identify all key hardware and software components as they relate to the interface.
>
> *Functional Allocation.* This subsection should describe the operations that are performed on each system involved in the interface. It should also describe how the end user would interact with the interface being defined. If the end user does not interact directly with the interface being defined, a description of the events that trigger the movement of information should be defined.
>
> *Data Transfer.* This subsection should describe how data would be moved among all component systems of the interface. Diagrams illustrating the connectivity among the systems are often helpful communication tools in support of this type of information.
>
> *Transactions.* This subsection should describe the types of transactions that move data among the component systems of the interface being defined. If multiple types of transactions are utilized for different portions of the interface, a separate section may be included for each interface.
>
> *Security and Safety.* If the interface defined has security and safety requirements, briefly describe how access security will be implemented and how data transmission security and safety requirements will be implemented for the interface being defined.

Detailed Interface Requirements. This section should provide a detailed description of all requirements in support of all interfaces between associated systems. This should include definitions of the content and format of every message or file that may pass between the two systems and the conditions under which each message or file is to be sent. The information presented in subsection Interface 1 Requirements should be replicated as needed to support the description of all interface requirements. Subsections are:

> *Interface 1 Requirements.* Briefly describe the interface, indicating data protocol, communication method(s), and processing priority.

Interface Processing Time Requirements. If the interface requires that data be formatted and communicated as the data is created, as a batch of data is created by operator action, or in accordance with some periodic schedule, indicate processing priority. Priority should be stated as measurable performance requirements, defining how quickly data requests must be processed by the interfacing system(s).

Message (or File) Requirements. This subsection should describe the transmission requirements. The definition, characteristics, and attributes of the requirements should be described.

Data Assembly Characteristics. This subsection should define all associated data elements that the interfacing entities must provide and the required access requirements.

Field/Element Definition. All characteristics of individual data elements should be described.

Communication Methods. This subsection should address all communication requirements, including a description of all connectivity and availability requirements. All aspects of the flow of communication should be described.

Security Requirements. This subsection should address all security features that are required in support of the interface process.

Note. When more than one interface between two systems is being defined in a single ICD, each should be defined separately, including all of the characteristics described in Interface 1 Requirements for each. There is no limit to the number of unique interfaces that can be defined in a single interface control document. In general, all interfaces defined should involve the same two systems.

Qualification Methods. This section should describe all qualification methods to be used to verify that the requirements for the interfaces have been met. Qualification methods may include:

- Demonstration. The operation of interfacing entities that relies on observable functional operation and does not require the use of instrumentation, special test equipment, or subsequent analysis.
- Test. The operation of interfacing entities using instrumentation or special test equipment to collect data for later analysis.
- Analysis. The processing of accumulated data obtained from other qualification methods. Examples are reduction, interpretation, or extrapolation of test results.
- Inspection. The visual examination of interfacing entities, documentation, and so on
- Special qualification methods. Any special qualification methods for the interfacing entities, such as special tools, techniques, procedures, facilities, and acceptance limits.

If a separate test plan exists, then a reference to this document should be provided.

System Integration Test Plan

The following information is based on IEEE Std 829-1998, IEEE Standard for Software Test Documentation [5], and IEEE Std 12207.0, Standard for Information Technology—

Life cycle processes [39] and has been adapted to support CMMI requirements. Additional information is provided in the document template, *System Integration Test Plan.doc,* which is located on the CD-ROM accompanying this book. Table 6-11 shows an outline of suggested document content.

The following provides section-by-section guidance in support of the creation of a system integration test plan. This guidance should be used to help define associated testing requirements and should reflect the actual requirements of the implementing organization.

Table 6-11. System integration test plan document outline

Title Page
Revision Page
Table of Contents
  1. Introduction
  2. Scope
    2.1 Identification and Purpose
    2.2 System Overview
    2.3 Definitions, Acronyms, and Abbreviations
  3. Referenced Documents
  4. System Integration Test Objectives
  5. System Integration Testing Kinds and Approaches
    5.1 Software Integration Testing
    5.2 COTS Integration Testing
    5.3 Database Integration Testing
    5.4 Hardware Integration Testing
    5.5 Prototype Usability Testing
    5.6 Approaches
  6. Development Test and Evaluation
    6.1 Development Test and Evaluation
      6.1.1 System and Software Items
      6.1.2 Hardware and Firmware
      6.1.3 Other Materials
      6.1.4 Proprietary Nature, Acquirer's Rights, and Licensing
      6.1.5 Participating Organizations
    6.2 Test Sites
  7. Test Identification
    7.1 General Information
      7.1.1 Test Levels
      7.1.2 Test Classes
      7.1.3 General Test Conditions
      7.1.4 Test Progression
    7.2 Planned Testing
      7.2.1 Integration Test
  8. Test Schedules
  9. Risk Management
 10. Requirements Traceability
 11. Notes
Appendix A. System Test Requirements Matrix
Appendix B. System Integration Test Description

Introduction. This section should provide a brief introductory overview of the project and related testing activities described in this document.

Scope. This section contains two subsections. The *Identification and Purpose* subsection should provide information that uniquely identifies the System effort and this associated test plan. This can also be provided in the form of a unique test plan identifier. The following text provides an example:

> This System Integration Test Plan details the testing planned for the [Project Name] ([Project Abbreviation]) Version [xx], Statement of Work, [date], Task Order [to number], Contract No. [Contract #], and Amendments.
> The goal of [project acronym] development is to [goal]. This System will allow [purpose].
> Specifics regarding the implementation of these modules are identified in the [Project Abbreviation] System Requirements Specification (SysRS) with line item descriptions in the accompanying Requirements Traceability Matrix (RTM) and references to the Interface Control Documentation (ICD).

The *System Overview* subsection should provide a summary of all System items and System features to be tested. The need for each item and its history may be addressed here as well. References to associated project documents should be cited here. The following provides a example:

> This document describes the System Integration Test Plan (SysITP) for the [PROJECT ABBREVIATION] System. [PROJECT ABBREVIATION] documentation will include this SysITP, the [PROJECT ABBREVIATION] System Requirements Specification (SysRS), the System Requirements Traceability Matrix (RTM), the Interface Control Documentation (ICD), the [PROJECT ABBREVIATION] Software Development Plan (SDP), the Software Design Document (SDD), the [PROJECT ABBREVIATION] User's Manual, the [PROJECT ABBREVIATION] System Administrator's Manual, and the [PROJECT ABBREVIATION] Data Dictionary.
> This SysITP describes the process to be used and the deliverables to be generated in the testing of the [PROJECT ABBREVIATION]. This plan, and the items defined herein, will comply with the procedures defined in the [PROJECT ABBREVIATION] Software Configuration Management (SCM) Plan and the [PROJECT ABBREVIATION] Software Quality Assurance (SQA) Plan. Any nonconformance to these plans will be documented as such in this SysITP.
> This document is based on the [PROJECT ABBREVIATION] System Integration Test Plan template with tailoring appropriate to the processes associated with the creation of the [PROJECT ABBREVIATION]. The information contained in this SysITP has been created for [Customer Name] and is to be considered "For Official Use Only."
> Please refer to Section "6. Schedule" in the SDP for information regarding documentation releases.

Definitions, Acronyms, and Abbreviations. All relevant definitions, acronyms, and abbreviations should be included in this section.

Referenced Documents. This section should include all material referenced during the creation of the test plan.

System Integration Test Objectives. Test objectives should include the verification of the product integration (i.e., to determine if it fulfills its system requirements specification and architectural design), the identification of defects that are not efficiently identified during unit testing, and the determination of the extent to which the subsystems/system is ready for system test. This section should provide project status metrics (e.g., percentage of test scripts successfully tested).

System Integration Testing Kinds and Approaches. Subsections are:

Software Integration Testing. Incremental integration testing of two or more integrated software components on a single platform or multiple platforms is conducted to produce failures caused by interface defects. This section should provide references to the SysRS, design documentation, and interface control document to show a complete list of subsystem or system modules and resulting test strategy.

COTS Integration Testing. Incremental integration testing of multiple commercial-off-the-shelf (COTS) software components is conducted to determine if they are not interoperable (i.e., if they contain any interface defects). This section should provide references to the SysRS, design documentation, and interface control document to show a complete list of subsystem or system modules and resulting test strategy.

Database Integration Testing. Incremental integration testing of two or more integrated software components is conducted to determine if the application software components interface properly with the database(s). This section should provide references to the SysRS, design documentation, and interface control document to show a complete list of subsystem or system modules and resulting test strategy.

Hardware Integration Testing. Incremental integration testing of two or more integrated hardware components in a single environment is conducted to produce failures caused by interface defects. This section should provide references to the SysRS, Design documentation, and Interface control document to show a complete list of subsystem or system modules and resulting test strategy.

Prototype Usability Testing. Incremental integration testing of a user interface prototype against its usability requirements is conducted to determine if it contains any usability defects. This section should provide references to the SysRS, Design documentation, and User's Manual to show a complete list of subsystem or system modules and resulting test strategy.

Approaches. Top-down testing consists of high-level components of a system that are integrated and tested before their design and implementation has been completed. Bottom-up testing consists of low level components that are integrated and tested before the high-level components have been developed. Object-oriented testing consists of use-case or scenario-based testing, thread testing, and/or object interaction testing. Interface testing consists of modules or subsystems that are integrated to create large systems.

System Integration Test Environment. This section contains the following six subsections.

Development Test and Evaluation. This subsection should provide a high-level summary of all key players, facilities required, and site of test performance. The following is provided as an example:

System Integration Tests are to be performed at [company name], [site location]. All integration testing will be conducted in the development center [room number]. The following individuals must be in attendance: [provide list of performers and observers].

System and Software Items. This subsection should provide a complete list of all System and software items used in support of integration testing, including the versions. If the system is kept in an online repository, reference to the storage location may be cited instead of listing all items here. The following is an example:

System and software used in the integration testing of [PROJECT ABBREVIATION]: [System and software List]. For details of client and server System specifications see the [PROJECT ABBREVIATION] System Requirements Specifications (SysRS).

Hardware and Firmware Items. This subsection should provide a complete list of all hardware and firmware items used in support of integration testing. If this information is available in another project document, it may be referenced instead of listing all equipment. The following is an example:

[Hardware Items] are to be used during integration testing, connected together in a client server relationship through a TCP/IP compatible network. For details of client and server hardware specifications, see the [PROJECT ABBREVIATION] System Requirements Specifications (SysRS). For integration testing, a representative hardware set as specified in the SysRS is to be used.

Other Materials. This subsection should list all other materials required to support integration testing.

Proprietary Nature, Acquirer's Rights, and Licensing. Any of the issues associated with the potential proprietary nature of the software, acquirer's rights, or licensing should be addressed in this subsection. An example is provided:

Licensing of commercial software is one purchased copy for each PC it is to be used on with the exception of [group software] which also requires licensing for the number of users/client systems. Some of the information in the data set is covered by the Privacy Act and will have to be protected in some manner.

Participating Organizations. This section should identify all groups responsible for integration testing. An example is provided:

The participating organizations are [Company Name], [customer name], and organizations at the beta test site.

These groups may participate in completing the Test Plan Walk-through Checklist. An example is identified in Appendix C, Software Process Work Products, of this text.

Test Site(s). This section should provide a description of test sites. See below:

The planned integration test sites are listed as follows: [Integration Sites].

Test Identification, General Information. This section contains the following four subsecitons.

Test Levels. This subsection should describe the different levels of testing required in support of the development effort. See Appendix C, Test Design Specification, Test Case Specification, and Test Procedure Specification sections. The following is provided as an example:

> Tests are to be performed at the integration level prior to release for system testing. Please refer to test design specifications #xx through xx.

Test Classes. This subsection should provide a description of all test classes. A test case specification should be associated with each of these classes. A summary of the validation method, data to be recorded, data analysis activity, assumptions, should be provided for each case.

General Test Conditions. This subsection should describe the anticipated baseline test environment. An example is provided below:

> A sample real data set from an existing database is to be used during all tests. The sample real data set will be controlled under the configuration management system. A copy of the controlled data set is to be used in the performance of all testing.

Test Progression. This subsection should provide a description of the progression of testing. A diagram is often helpful when attempting to describe the progression of testing. The information below is provided as an example:

> The System Integration Testing is test verifying that all subsystems or modules work together. The module tests are performed prior or during the Implementation phase as elements and modules are completed. All module and integration tests must be passed before performing System Testing. The Change Enhancement Request (CER) tracking system is used to determine eligibility for testing at a level. See the [PROJECT ABBREVIATION] Software Configuration Management Plan and [PROJECT ABBREVIATION] Software Quality Assurance Plan for details on CER forms and processes.

Planned Testing. This section should provide a detailed description of the type of testing to be employed. The following is provided as an example:

> A summary of testing is provided in section xxx. Additional information is provided in Appendix A and B.
>
> *Integration Test.* All of the modules to be integration tested (refer to System Requirements Traceability Matrix) will be tested using integration-level test methodology. For details of the procedures and setup see the [PROJECT ABBREVIATION] System Integration Test Description. The resulting outputs of this test are Internal Test Reports (ITR) or System Integration Test Report (SITR). When these integration tests are all passed the [PROJECT ABBREVIATION] System will be ready for system level testing.
>
> For the integration level tests the following classes of tests will be used:
> User interaction behavior consistency
> Display screen and printing format consistency
> Check interactions between modules
> Measure time of reaction to user input

Test Schedules. This section should include test milestones as identified in the system project schedule as well as any required deliverables associated with testing. Any additional milestones may be defined here as needed. If appropriate detail is provided in the software project management plan (SRMP), then this can be referenced. The following is provided as an example:

> Test schedules are defined in the [PROJECT ABBREVIATION] Software Development Plan.

Risk Management. This section should identify all high-risk assumptions of the test plan. The contingency plans for each should be described and the management of risk items should be addressed.

Requirements Traceability. This section should provide information regarding the traceability of testing to the requirements and design of the system. A matrix, or database, is useful when meeting this traceability requirement. The following is provided as an example:

> Refer to [PROJECT ABBREVIATION] System Requirements Specification, Appendix A for information regarding requirements traceability.

Note. This section should provide any additional information not previously covered in the test development plan.

Appendix A. System Test Requirements Matrix. This is redundant. Only refer to SysRS Appendix A and attach test information as another column.

Appendix B. System Integration Test Description. This should contain a detailed description of each test.

VERIFICATION

The Verification process area addresses the preparation, performance, and identification of corrective actions associated with verification activities. These activities include the verification of the intermediate and final product against all requirements. Verification should occur throughout the software development life cycle, concluding with the verification of a completed product.

IEEE Std 12207.0, Standard for Information Technology—Software Life Cycle Processes [39], describes 17 processes spanning the entire life cycle of a software product or service. Even if an organization's processes were defined using other sources, the standard is useful in characterizing the essential characteristics of these software processes and should be considered prior to the implementation of process improvement activities. Referencing this standard and reviewing what is required for each of these primary process areas can provide additional guidance in support of the activities associated with verification activities. It is important that IEEE 12207 be considered prior to the implementation of any process improvement activities in support of the verification PA.

CMMI®-SW Goals

The activities in support of this process area should identify all items requiring verification, describe the methods used to perform the verification, and associate all requirements. Verification helps to ensure that the product will meet customer requirements.

SG1. Prepare for Verification
Preparation for verification is conducted.

Preparation is critical in order to ensure that verification activities are associated with all requirements, designs, developmental plans, and schedules. Methods of verification can include inspections, peer reviews, audits, walk-throughs, analyses, simulations, testing, and demonstrations. All associated support tools, test equipment and software, simulations, prototypes, and facilities must be identified.

SG2. Perform Peer Reviews
Peer reviews are performed on selected work products.

Peer-driven inspections, structured walk-throughs, or other peer reviews are an effective way to identify defects in work products. These reviews should not only be applied to work products but also to supporting process documentation, tools, and material.

SG3. Verify Selected Work Products
Selected work products are verified against their specified requirements.

Verifying products promotes the early detection of problems, resulting in the early removal of defects. These results of verification save considerable cost and time. For users of the continuous representation, this is a capability-Level-1-specific practice. When there are no procedures or criteria established, use the methods established by the Select Work Products for Verification specific practice to accomplish capability Level 1 performance [61].

GG 3. Institutionalize a Defined Process
The process is institutionalized as a defined process.

Peer reviews are an important part of verification and are a proven mechanism for effective defect removal. Peer reviews are often conducted in the form of an inspection or walk-through. In addition to defect removal, peer reviews provide insight into the processes behind the development of specific work products. These verification activities not only effectively help to identify and remove defects, but also help to prevent them. The delivery and documentation associated with inspections and walk-throughs are discussed in the following. Additional associated work product information is provided in Appendix C, Inspection Report Description, and Inspection Log Defect Summary Description.

Related PAs include the Validation process area, which provides additional to help ensure that a specified work product fulfills its intended use; the Requirements Development process area, providing additional information about the generation and development of customer, product, and product-component requirements; and the Requirements Management process area that provides additional information about the effective management of requirements. Table 6-12 describes the specific and generic goals and practices, typical work products, and commonly associated plans or artifacts in support of the PA.

Inspections

The information provided here is based upon the recommendations provided in IEEE Std 1028-1997, IEEE Standard for Software Reviews. Inspections are an effective way to de-

Table 6-12. Verification goals and practices [61]

Specific and generic goals/practices and typical work products	Plan or artifact
SG1. Prepare for Verification	
SP 1.1. Select Work Products for Verification	
Lists of work products selected for verification	Software project management plan
Verification methods for each selected work product	Quality assurance plan
SP 1.2. Establish the Verification Environment	
Verification environment	Quality assurance plan
SP 1.3. Establish Verification Procedures and Criteria	
Verification procedures and criteria	Quality assurance plan
SG2. Perform Peer Reviews	
SP 2.1. Prepare for peer reviews	
Peer review schedule and checklist	Software project management plan and quality assurance plan
Entry and exit criteria for work products	Software project management plan and quality assurance plan
Criteria for requiring another peer review	Software project management plan and quality assurance plan
Selected work products to be reviewed	Software project management plan and quality assurance plan
SP 2.2. Conduct Peer Reviews	
Peer review results, issues, data	Software project management plan (schedule) and quality assurance plan, inspection log defect summary description, inspection report description, SQA inspection log
SP 2.3. Analyze Peer Review Data	
Peer review data	Quality assurance plan
Peer review action items	Quality assurance plan
SG3. Verify Selected Work Products	
SP 3.1. Perform Verification	
Verification results and reports	Quality assurance plan
Demonstrations	Quality assurance plan
As-run procedures log	Quality assurance plan
SP 3.2. Analyze Verification Results and Identify Corrective Action	
Analysis report	Quality assurance plan
Trouble reports	Quality assurance plan
Change requests	Quality assurance plan
Corrective actions to verification methods, criteria, and/or environment	Quality assurance plan

Table 6-12. *Continued*

Specific and generic goals/practices and typical work products	Plan or artifact
GG3. Institutionalize a Defined Process	
GP 2.1. Establish an organizational policy	
Establishes organizational expectations for establishing and maintaining verification methods, procedures, criteria, verification environment, performing peer reviews, and verifying selected work products	Organizational policy
GP 2.2. Plan the process	
Typically, this plan for performing the verification process is included in (or referenced by) the project plan, which is described in the Project Planning process area	Software project management plan and quality assurance plan
GP 2.3. Provide resources	
Identification of resources used in support of verification activities	Software project management plan and quality assurance plan
GP 2.4. Assign responsibility	Software project management plan
GP 2.5. Train people	Software project management plan or training plan
GP 2.6. Manage configurations	
Description of how work products and all associated artifacts in support of verification activities are placed under configuration management	Configuration management plan
GP 2.7. Identify relevant stakeholders	
Describe activities and feedback mechanisms	Software project management plan
GP 2.8. Monitor and control the process	Software project management or software measurement and metrics plans
GP 2.9. Objectively evaluate adherence	Quality assurance plan
GP 2.10. Review status with higher-level management	Software project management plan
GP 3.1. Establish a defined process	
Establish and maintain a description of the verification process	Quality assurance plan
GP 3.2. Collect improvement information	Software measurement and metrics plan, or software project management plan

tect and identify problems early in the software development process. Inspections should be peer-driven and should follow a defined, predetermine practice. The criteria for inspections are presented here to help organization define their inspection practices. IEEE Std 1044-1993, IEEE Standard Classification for Software Anomalies[13] provides valuable information regarding the categorization of software anomalies. Additional information is provided in Appendix C, which presents a recommended minimum set of software reviews and an example SQA inspection log.

Introduction. Software inspections can be used to verify that a specified software product meets its requirements. These requirements can address functional characteristics, quality attributes, or adherence to regulations or standards. Any inspection practice should collect anomaly data that can be used to identify trends and improve the development process. Use the Classic Anomaly Class Categories section in Appendix C, Software Process Work Products.

Examples of software products subject to inspections are:

a. Software requirements specification
b. Software design description
c. Source code
d. Software test plan documentation
e. Software user documentation
f. Maintenance manual
g. System build procedures
h. Installation procedures
i. Release notes

Responsibilities. Each inspection should have a predefined set of roles and responsibilities; the assignment of these responsibilities should occur on a case-by-case basis and is dependent upon the item subject to inspection.

Participants. Inspections should consist of three to six participants and should be led by an impartial facilitator who is trained in inspection techniques. Determination of remedial or investigative action for an anomaly is a mandatory element of a software inspection, although the resolution should not occur in the inspection meeting. Collection of data for the purpose of analysis and improvement of software engineering procedures (including all review procedures) is strongly recommended but is not a mandatory part of software inspections. The participants are:

Inspection leader. The inspection leader is responsible for administrative tasks pertaining to the inspection and is responsible for the planning and preparation of the inspection. The leader is responsible for ensuring that the inspection is conducted in an orderly manner and meets its objectives. The leader is responsible for the collection of inspection data and the issuance of the inspection report.

Recorder. The recorder is responsible for recording all inspection data required for process analysis. The recorder should document all anomalies, action items, decisions, and recommendations made by the inspection team. The inspection leader may also act as recorder.

Reader. The reader should methodically lead the inspection team through the software product, providing a summary, when appropriate, highlighting all items of interest.

Author. The author contributes the product for inspection and participates when a special understanding of the software product is required. The author is responsible for any rework required to make the software product meet inspection exit criteria.

Inspector. All participants in the review are inspectors. The author should not act as inspection leader and should not act as reader or recorder. All other roles may be shared among the team members and participants may act in more than one role. Individuals holding management positions over any member of the inspection team should not participate in the inspection.

Inspectors are responsible for the identification and description of all anomalies found in the software product. Inspectors should represent different viewpoints at the meeting (for example, sponsor, requirements, design, code, safety, test, independent test, project management, quality management, and hardware engineering). Only those viewpoints with relevance to the product should be present.

Each inspector should be assigned a specific review topic to ensure effective product coverage. The inspection leader should assign these roles prior to the inspection.

Input. Input to the inspection should include the objectives, identification of the software product, the inspection procedure, all reporting forms, all guidelines or standards, and any known anomalies or issues.

Authorization. Inspections should be identified in any associated software project management plan or software verification and validation plan, and should be reflected in project scheduling and resource allocation.

Preconditions. An inspection should be conducted only when a statement of objectives for the inspection is established and the required inspection inputs are available.

Minimum Entry Criteria. An inspection should not be conducted until the software product is complete and conforms to project standards for content and format, all tools and documentation required in support of the inspection are available, and any related prior milestones have been satisfied. If the inspection is a reinspection, all previously identified problems should have been resolved.

Management Preparation. Managers are responsible for planning the time and resources required for inspection, including support functions, and ensuring that these are identified in the associated software project management plan. Managers should also ensure that individuals receive adequate training and orientation on inspection procedures and possess appropriate levels of expertise and knowledge sufficient to comprehend the software product under inspection.

Planning the Inspection. The author of the software product is responsible for the assembly of all inspection materials. The inspection leader is responsible for the identification of the inspection team, the assignment of specific responsibility, and scheduling and conducting the inspection. As a part of the planning procedure, the inspection team should determine if anomaly resolution is to occur during the inspection meeting or if resolution will occur under other circumstances.

Overview of Inspection Procedures. The author of the software product should present a summary overview for the inspectors. The inspection leader should answer questions about any checklists and assign roles. Expectation should be reviewed, such as the minimum preparation time anticipated and the typical number of anticipated anomalies.

Preparation. Each inspection team member should examine the software product and document all anomalies found. All items found should be sent to the inspection leader. The inspection leader should classify (i.e., consolidate) all anomalies and forward these to the author of the software product for disposition. The inspection leader, or reader, should specify the order of the inspection (e.g., hierarchical). The reader should be prepared to present the software product at the inspection meeting.

Examination. The inspection meeting should follow this agenda:

Introduce the meeting. The inspection leader should introduce all participants and review their roles. The inspection leader should state the purpose of the inspection and should remind the inspectors to focus their efforts toward anomaly detection, direct remarks to the reader and not the author, and to comment only on the software product and not the author. See Classic Anomaly Class Categories in Appendix C.

Establish preparedness. The inspection leader should verify that all participants are prepared. The inspection leader should gather individual preparation times and record the total in associated inspection documentation.

Review general items. General, nonspecific, or pervasive anomalies should be presented first. If problems are discovered that inhibit the further efficient detection of error, the leader may choose to reconvene the inspection following the resolution of all items found.

Review software product and record anomalies. The reader should present the software product to the inspection team. The inspection team should examine the software product and create a list of anomalies found. The recorder should enter each anomaly, location, description, and classification on the anomaly list.* See Inspection Log Defect Summary Description in Appendix C.

Review the anomaly list. At the end of the inspection meeting, the team should review the anomaly list to ensure its completeness and accuracy. The inspection leader should allow time to discuss every anomaly about which any disagreement occurred.

Make an exit decision. The purpose of the exit decision is to bring an unambiguous closure to the inspection meeting. Common exit criteria are:

a. *Accept with no or minor rework.* The software product is accepted as is or with only minor rework (requires no further verification).

b. *Accept with rework verification.* The software product is to be accepted after the inspection leader or a designated member of the inspection team verifies rework.

c. *Reinspect.* Schedule a reinspection to verify rework. At a minimum, a reinspection should examine the software product areas changed to resolve anomalies identified in the last inspection, as well as side effects of those changes.

Exit Criteria. An inspection is considered complete when prescribed exit criteria have been met.

Output. The output of the inspection provides documented evidence that identifies that the project is being inspected. It provides a list of all team members, the meeting duration,

---

*IEEE Std 1044 [13] may be used to support anomaly classification.

a description of the product inspected, all inspection inputs, objectives, anomaly list and summary, a rework estimate, and total inspection preparation time.

Data Collection Recommendations. Inspection data should contain the identification of the software product, the date and time of the inspection, the inspection leader, the preparation and inspection times, the volume of the materials inspected, and the disposition of the inspected software product. The management of inspection data should provide for the capability to store, enter, access, update, summarize, and report categorized anomalies.

Anomaly Classification. Anomalies may be classified by technical type. IEEE Std 1044-1993 provides effective guidance in support of anomaly classification. Appendix A of IEEE Std 1044 provides sample screens from an anomaly reporting system. The data fields for this system are listed in Appendix A under the Verification Artifacts section. Also see Classic Anomaly Class Categories in Appendix C of this book.

Anomaly Ranking. Anomalies may be ranked by potential impact on the software product. Three ranking categories are provided as examples:

1. *Category 1.* Major anomaly of the software product, or an observable departure from specification, with no available workaround.
2. *Category 2.* Minor anomaly of the software product, or departure from specification, with existing workaround.
3. *Category 3.* Cosmetic anomalies that deviate from relevant specifications but do not cause failure of the software product.

The information is captured in the Inspection Report Description and the Inspection Log Defect Summary Description as shown in Appendix C.

Improvement. Inspection data should be analyzed regularly in order to improve the inspection and software development processes. Frequently occurring anomalies may be included in the inspection checklists or role assignments. The checklists themselves should also be evaluated for relevance and effectiveness.

Walk-throughs

The purpose of a systematic walk-through is to evaluate a software product. Walk-throughs effectively identify anomalies and contribute to the improvement of the software development process. Walk-throughs are also effective training activities, encouraging technical exchanges and information sharing. Examples of software products subject to walk-throughs include software project documentation, source code, and system build or installation procedures.

Responsibilities. The following roles should be established:

Leader. The walk-through leader is responsible for administrative tasks pertaining to the walk-through and is responsible for the planning and preparation of the walk-through. The leader is responsible for ensuring that the walk-through is conducted in an orderly manner and meets its objectives. The leader is responsible for the collection of walk-through data and the issuance of the summary report.

Recorder. The recorder should methodically document all decisions, actions, and anomalies discussed during the walk-through.

Author. The author contributes the product for walk-through and participates when a special understanding of the software product is required. The author is responsible for any rework required to address all anomalies identified.

Team Member. All participants in the walk-through are team members. All other roles may be shared among the team members and participants may act in more than one role. Individuals holding management positions over any member of the walk-through team shall not participate in the walk-through.

Input. Input criteria should include the walk-through objectives, the software product, anomaly categories,* and all relevant standards. See Classic Anomaly Class Categories in Appendix C.

Authorization. Walk-throughs should be identified in any associated software project management plan, software verification and validation plan, and should be reflected in project scheduling and resource allocation.

Preconditions. A walk-through should be conducted only when a statement of objectives for the walk-through is established and the required walk-through inputs are available.

Management Preparation. Managers are responsible for planning the time and resources required for the walkthrough, including support functions, and that these are identified in the associated software project management plan. Managers should also ensure that individuals receive adequate training and orientation on walkthrough procedures and possess appropriate levels of expertise and knowledge sufficient to evaluate the software product.

Planning the Walk-through. The walk-through leader is responsible for identifying the walk-through team, distributing walk-through materials, and scheduling and conducting the walk-through.

Overview. The author of the software product should present a summary overview for the team.

Preparation. Each walk-through team member should examine the software product and prepare a list of items for discussion. These items should be categorized as either specific or general, where "general" would apply to the entire product and specific to one part. All anomalies detected by team members should be sent to the walk-through leader for classification. The leader should forward these to the author of the software product for disposition. The walk-through leader should specify the order of the walk-through (e.g., hierarchical).

Examination. The walk-through leader should introduce all participants, describe their roles, and state the purpose of the walk-through. All team members should be reminded to

---

*IEEE Std 1044 [13] may be used to support anomaly classification.

focus their efforts toward anomaly detection. The walk-through leader should remind the team members to comment only on the software product and not its author. Team members may pose general questions to the author regarding the software product.

The author should then present an overview of the software product under review. This is followed by a general discussion during which team members raise their general concerns. After the general discussion, the author serially presents the software product in detail. Team members raise their specific concerns when the author reaches them in the presentation. It is important to note that new items of concern may be raised during the meeting.

The walk-through leader is responsible for guiding the meeting to a decision or action on each item. The recorder is responsible for noting all recommendations and required actions. Appendix C, Software Process Work Products, contains several walk-through forms: Requirements Walk-through Form, Software Project Plan Walk-through Checklist, Preliminary Design Walk-through Checklist, Detailed Design Walk-through Checklist, Program Code Walk-through Checklist, Test Plan Walk-through Checklist, Walk-through Summary Report, and Classic Anomaly Class Categories.

After the walk-through meeting, the walk-through leader should issue a walk-through report detailing anomalies, decisions, actions, and other information of interest. This report should include a list of all team members, description of the software product, statement of walk-through objectives, list of anomalies and associated actions, all due dates and individuals responsible, and the identification of follow-up activities.

Data Collection Recommendations. Walk-through data should contain the identification of the software product, the date and time of the walk-through, the walk-through leader, the preparation and walk-through times, the volume of the materials inspected, and the disposition of the software product. The management of walk-through data should provide for the capability to store, enter, access, update, summarize, and report categorized anomalies.

Anomaly Classification. Anomalies may be classified by technical type. IEEE Std 1044-1993 provides effective guidance in support of anomaly classification. Appendix A of IEEE Std 1044 provides sample screens from an anomaly reporting system. The data fields for this system are listed in Appendix A under the Verification Artifacts section.

Anomaly Ranking. Anomalies may be ranked by potential impact on the software product. Three ranking categories are provided as examples:

1. *Category 1.* Major anomaly of the software product, or an observable departure from specification, with no available workaround.
1. *Category 2.* Minor anomaly of the software product, or departure from specification, with existing workaround.
1. *Category 3.* Cosmetic anomalies that deviate from relevant specifications but do not cause failure of the software product.

Improvement. Walk-through data should be analyzed regularly in order to improve the walk-through and software development processes. Frequently occurring anomalies may be included in the walk-through checklists or role assignments. The checklists themselves should also be evaluated for relevance and effectiveness.

VALIDATION

The Verification and Validation process areas are similar, but they address different issues. Validation demonstrates that the product will fulfill its intended use, whereas verification addresses whether the work product meets specified requirements. In other words, verification ensures that "you built it right," whereas validation ensures that "you built the right thing."

It is important to remember that validation should be accomplished in the target environment. The entire environment may be used or only part of it. However, validation issues can be discovered more effectively and earlier if the validation reflects deployed conditions.

CMMI®-SW Goals

SG 1. Prepare for Validation
Preparation for validation is conducted.

Preparation is critical in order to ensure that validation activities are associated with all products and product components selected for validation and establishing and maintaining the validation environment, procedures, and criteria. All associated support tools, test equipment and software, simulations, prototypes, and facilities must be identified.

SG 2. Validate Product or Product Components
The product or product components are validated to ensure that they are suitable for use in their intended operating environment.

This specific goal aims to ensure that all identified validation methods, procedures, and defined criteria are used to validate selected products.

GG 3. Institutionalize a Defined Process
The process is institutionalized as a defined process.

The Requirements Development process area provides additional information about requirements validation.

The Technical Solution process area provides more information about converting requirements into product specifications and corrective action when validation issues are identified.

The Verification process area provides additional information in support of verifying that the product or product component meets its requirements. Table 6-13 describes the specific and generic goals and practices, typical work products, and commonly associated plans or artifacts in support of the PA.

Software Test Plan

The documentation supporting software testing should define the scope, approach, resources, and schedule of the testing activities. It should identify all items being tested, the features to be tested, the testing tasks to be performed, the personnel responsible for each task, and the risks associated with testing. IEEE Std 829-1998, IEEE Standard for Soft-

Table 6-13. Validation goals and practices [61]

Specific and generic goals/practices and typical work products	Plan or artifact
SG1. Prepare for Validation	
SP 1.1. Select Products for Validation	
Lists of products and product components selected for validation	Software test plan
Validation methods for each product or product component	Software test plan
Requirements for performing validation for each product or product component	Requirements specification or tracking database
Validation constraints for each product or product component	Software test plan
SP 1.2. Establish the Validation Environment	
Validation environment	Software test plan
SP 1.3. Establish Validation Procedures and Criteria	
Validation procedures and criteria	Software test plan
Test and evaluation procedures for maintenance, training, and support	Software test plan
SG2. Validate Product or Product Components	
SP 2.1. Perform Validation	
Validation reports, results, cross-reference matrix	Software test plan and requirements tracking database
As-run procedures log	Software test plan, test log
Operational demonstrations	Software test plan
SP 2.2. Analyze Validation Results	
Validation deficiency reports	Software test plan, test incident report
Validation issues	Software test plan, test incident report, test summary report
Procedure change request	Software test plan, requirements management plan, or CCB charter
GG3. Institutionalize a Defined Process	
GP 2.1. Establish an organizational policy	
Establishes organizational expectations for establishing and maintaining verification methods, procedures, criteria, verification environment, performing peer reviews, and verifying selected work products	Organizational policy
GP 2.2. Plan the process	Software project management plan
GP 2.3. Provide resources	
Identification of resources used in support of validation activities	Software project management plan
GP 2.4. Assign responsibility	Software project management plan
GP 2.5. Train people	Software project management plan or training plan

(continued)

Table 6-13. *Continued*

Specific and generic goals/practices and typical work products	Plan or artifact
GG3. Institutionalize a Defined Process (*cont.*)	
GP 2.6. Manage configurations	
Description of how work products and all associated artifacts in support of validation activities are placed under configuration management	Configuration management plan
GP 2.7. Identify relevant stakeholders	
Describe activities and feedback mechanisms	Quality assurance plan
GP 2.8. Monitor and control the process	Software project management or software measurement and metrics plans
GP 2.9. Objectively evaluate adherence	Quality assurance plan
GP 2.10. Review status with higher-level management	Software project management plan
GP 3.1. Establish a defined process	
Establish and maintain a description of the validation process	Quality assurance plan
GP 3.2. Collect improvement information	Software measurement and metrics plan, or software project management plan

ware Test Documentation [5] and IEEE 12207.1—Standard for Information Technology—Software Life Cycle Processes—Life Cycle Data were used as primary references for the information provided here.

The IEEE Computer Society also publishes IEEE Std 1008-1997 (R2003), IEEE Standard for Software Unit Testing [8], and IEEE Std 1012-1998, IEEE Standard for Software Verification and Validation [9] as part of their software engineering standards collection. These documents provide additional information in support of the development and definition of software test processes and procedures. Table 6-14 provides an example document outline. Additional information in support of the work products commonly associated with this PA may be found in Appendix C.

Software Test Plan Document Guidance

The following provides section-by-section guidance in support of the creation of a software test plan. It does not identify specific testing methodologies, approaches, techniques, facilities, or tools. As with any defined software engineering process, additional supporting documentation may be required. The development of Software Test Plan (STP) is integral to the software development process. Additional information is provided in the document template, *Software Test Plan.doc,* which is located on the CD-ROM accompanying this book.

Identification. This section should provide information that uniquely identifies the software effort and its associated test plan. This can also be provided in the form of a unique test plan identifier. The following text provides an example:

This software test plan is to detail the testing planned for the [Project Name] ([Project Abbreviation]) Version [xx], Statement of Work, [date], Task Order [to number], Contract No. [Contract #], and Amendments.

The goal of [project acronym] development is to [goal]. This software will allow [purpose].

Specifics regarding the implementation of these modules are identified in the [Project Abbreviation] Software Requirements Specification (SRS) with line item descriptions in the accompanying Requirements Traceability Matrix (RTM).

Document Overview. This section should provide a summary of all software items and software features to be tested. The need for each item and its history may be addressed here as well. References to associated project documents should be cited here. The following provides an example:

This document describes the Software Test Plan (STP) for the [PROJECT ABBREVIATION] software. [PROJECT ABBREVIATION] documentation will include this STP, the [PROJECT ABBREVIATION] Software Requirements Specification (SRS), the Requirements Traceability Matrix (RTM), the [PROJECT ABBREVIATION] Software Development Plan (SDP), the Software Design Document (SDD), the [PROJECT ABBREVIATION] User's Manual, the [PROJECT ABBREVIATION] System Administrator's Manual, and the [PROJECT ABBREVIATION] Data Dictionary.

This STP describes the process to be used and the deliverables to be generated in the testing of the [PROJECT ABBREVIATION]. This plan, and the items defined herein, will comply with the procedures defined in the [PROJECT ABBREVIATION] Software Configuration Management (SCM) Plan and the [PROJECT ABBREVIATION] Software Quality Assurance (SQA) Plan. Any nonconformance to these plans will be documented as such in this STP.

This document is based on the [PROJECT ABBREVIATION] Software Test Plan template with tailoring appropriate to the processes associated with the creation of the [PROJECT ABBREVIATION]. The information contained in this STP has been created for [Customer Name] and is to be considered "For Official Use Only."

Please refer to Section "6. Schedule" in the SDP for information regarding documentation releases.

Acronyms and Definitions. All relevant acronyms and definitions should be included in this section.

Referenced Documents. This section should include all material referenced during the creation of the test plan.

Development Test and Evaluation. Provide a high-level summary of all key players, facilities required, and site of test performance. The following is provided as an example:

Qualification, integration, and module level tests are to be performed at [company name], [site location]. All testing will be conducted in the development center [room number]. The following individuals must be in attendance: [provide list of performers and observers].

Software Items. This section should provide a complete list of all software items used in support of testing, to include versions. If the software is kept in an online repository reference to the storage location may be cited instead of listing all items here. The following is an example:

Table 6-14. Software test plan document outline

Title Page
Revision Page
Table of Contents
1. Introduction
2. Scope
    2.1 Identification
    2.2 System Overview
    2.3 Document Overview
    2.4 Acronyms and Definitions
        2.4.1 Acronyms
        2.4.2 Definitions
3. Referenced Documents
4. Software Test Environment
    4.1 Development Test and Evaluation
        4.1.1 Software Items
        4.1.2 Hardware and Firmware Items
        4.1.3 Other Materials
        4.1.4 Proprietary Nature, Acquirer's Rights, and Licensing
        4.1.5 Installation, Testing, and Control
        4.1.6 Participating Organizations
    4.2 Test Sites
        4.2.1 Software Items
        4.2.2 Hardware and Firmware Items
        4.2.3 Other Materials
        4.2.4 Proprietary Nature, Acquirer's Rights, and Licensing
        4.2.5 Installation, Testing, and Control
        4.2.6 Participating Organizations
5. Test Identification
    5.1 General Information
        5.1.1 Test Levels
        5.1.2 Test Classes
            5.1.2.1 Check for Correct Handling of Erroneous Inputs
            5.1.2.2 Check for Maximum Capacity
            5.1.2.3 User Interaction Behavior Consistency
            5.1.2.4 Retrieving Data
            5.1.2.5 Saving Data
            5.1.2.6 Display Screen and Printing Format Consistency
            5.1.2.7 Check Interactions Between Modules
            5.1.2.8 Measure Time of Reaction to User Input
            5.1.2.9 Functional Flow
        5.1.3 General Test Conditions
        5.1.4 Test Progression
    5.2 Planned Tests
        5.2.1 Qualification Test
        5.2.2 Integration Tests
        5.2.3 Module Tests
        5.2.4 Installation Beta Tests
6. Test Schedules
7. Risk Management
8. Requirements Traceability
9. Notes
APPENDIX A. Software Test Requirements Matrix
APPENDIX B. Qualification Software Test Description
APPENDIX C. Integration Software Test Description
APPENDIX D. Module Software Test Description

Software used in the testing of [PROJECT ABBREVIATION]: [Software List]. For details of client and server software specifications see the [PROJECT ABBREVIATION] Software Requirements Specifications (SRS).

Hardware and Firmware Items. This section should provide a complete list of all hardware and firmware items used in support of testing. If this information is available in another project document, it may be referenced instead of listing all equipment. The following is an example:

> [Hardware Items] are to be used during testing, connected together in a client server relationship through a TCP/IP compatible network. For details of client and server hardware specifications see the [PROJECT ABBREVIATION] Software Requirements Specifications (SRS). For integration and qualification level testing a representative hardware set as specified in the SRS is to be used.

Other Materials. This section should list all other materials required to support testing.

Proprietary Nature, Acquirer's Rights, and Licensing. Any of the issues associated with the potential proprietary nature of the software, acquirer's rights, or licensing should be addressed in this section. An example is provided:

> Licensing of commercial software is one purchased copy for each PC it is to be used on with the exception of [group software], which also requires licensing for the number of users/client systems. Some of the information in the data set is covered by the Privacy Act and will have to be protected in some manner.

Installation, Testing, and Control. This section should address environment requirements or actions required to support the installation of the application for testing. The following is provided as an example:

> [Company Name] will acquire the software and hardware needed as per the [PROJECT ABBREVIATION] Software Development Plan (SDP) in order to run [PROJECT ABBREVIATION] in accordance with contract directions and limitations.
>
> [Company Name] will install the supporting software and test it per the procedures in the [PROJECT ABBREVIATION] Installation Software Test Description.
>
> The [PROJECT ABBREVIATION] Change Enhancement Request (CER) tracking system will be used to determine eligibility for testing at any level. See the [PROJECT ABBREVIATION] Software Configuration Management Plan and [PROJECT ABBREVIATION] Software Quality Assurance Plan for details on CER forms and processes.

Participating Organizations. This section should identify all groups responsible for testing. They may also participate in completing the Test Plan Walk-through Checklist (see Appendix C). An example is:

> Test Sites. This section should provide a description of test sites. See below.
> The planned installation level Beta test sites are listed as follows: [Installation Beta Sites]

Software Items. This section should provide a complete list of all software items used in support of site testing, including the versions. If the software is kept in an online repository, reference to the storage location may be cited instead of listing all items here. The following is an example:

Software used in the testing of [PROJECT ABBREVIATION]: [Software Items]. For details of client and server software specifications see the [PROJECT ABBREVIATION] Software Requirements Specifications (SRS).

Hardware and Firmware Items. This section should provide a complete list of all hardware and firmware items used in support of site testing. If this information is available in another project document, it may be referenced instead of listing all equipment. The following is an example:

IBM compatible PC's are to be used during testing, connected together in a client–server relationship through a TCP/IP compatible network. For details of client and server hardware specifications see the [PROJECT ABBREVIATION] Software Requirements Specifications (SRS). For integration and qualification-level testing a representative hardware set as specified in the SRS is to be used.

Other Materials. This section should list all other materials required to support testing.

Proprietary Nature, Acquirer's Rights, and Licensing. Any of the issues associated with the potential proprietary nature of the software, acquirer's rights, or licensing should be addressed in this section. An example is:

Licensing of commercial software is one bought copy for each PC it is to be used on with the exception of [group software], which also requires licensing for the number of users/client systems. Some of the information in the data set is covered by the Privacy Act and will have to be protected in some manner.

Installation, Testing, and Control. This section should address environment requirements or actions required to support the installation of the application for beta testing. The following is provided as an example:

The customer will acquire the software and hardware needed as defined in the [PROJECT ABBREVIATION] SRS and the SOW in order to run [PROJECT ABBREVIATION].

[Company Name] will install the supporting software and test it per the procedures in the [PROJECT ABBREVIATION] Installation Software Test Description and as defined in the [PROJECT ABBREVIATION] SDP.

The CER tracking system will be used to determine eligibility for testing at any level. See the [PROJECT ABBREVIATION] Software Configuration Management Plan and [PROJECT ABBREVIATION] Software Quality Assurance Plan for details on CER forms and processes.

Participating Organizations. This section should identify all groups responsible for testing. They may also participate in completing the Test Plan Walk-through Checklist. An example is:

The participating organizations are [Company Name], [customer name], and organizations at the beta test site.

Test Idenditification. General Information. Test Levels. This section should describe the different levels of testing required in support of the development effort. See Appendix C,

Test Design Specification, Test Case Specification, and Test Procedure Specification sections. The following is provided as an example:

> Tests are to be performed at the module, integration, installation, and qualification levels prior to release for beta testing. Please refer to test design specifications #xx through xx.

Test Classes. This section should provide a description of all test classes. A test case specification should be associated with each of these classes. A summary of the validation method, data to be recorded, data analysis activity, and assumptions should be provided for each case. An example is:

> Test class # xx. Check for correct handling of erroneous inputs.
>
> Test Objective. Check for proper handling of erroneous inputs: characters that are not valid for this field, too many characters, not enough characters, value too large, value too small, all selections for a selection list, no selections, all mouse buttons clicked or double clicked all over the client area of the item with focus.
>
> Validation Methods Used—Test.
>
> Recorded Data. User action or data entered, screen/view/dialog/control with focus, resulting action.
>
> Data Analysis. Was resulting action within general fault handling defined capabilities in the [PROJECT ABBREVIATION] SRS and design in [PROJECT ABBREVIATION] SDD?
>
> Assumptions and Constraints—None.

Additional examples are provided in support of test case development in *Software Test Plan.doc* on the CD-ROM accompanying this book.

General Test Conditions. This section should describe the anticipated baseline test environment. An example is:

> A sample real data set from an existing database is to be used during all tests. The sample real data set will be controlled under the configuration management system. A copy of the controlled data set is to be used in the performance of all testing.

Test Progression. This section should provide a description of the progression of testing as shown in Figure 6-1. A diagram is often helpful when attempting to describe the progression of testing. The information below is provided as an example:

> The Qualification Testing is a qualification-level test verifying that all requirements have been met. The module and integration tests are performed as part of the Implementation phase as elements and modules are completed. All module and integration tests must be passed before performing Qualification Testing. All module tests must be passed before performing associated integration-level tests. The CER tracking system is used to determine eligibility for testing at a level. See the [PROJECT ABBREVIATION] Software Configuration Management Plan and [PROJECT ABBREVIATION] Software Quality Assurance Plan for details on CER forms and processes.

Planned Testing. This section should provide a detailed description of the type of testing to be employed. The following is provided as an example:

> A summary of testing is provided in Section 4.3. Additional information is provided in Appendixes A, B, C, D, and E.

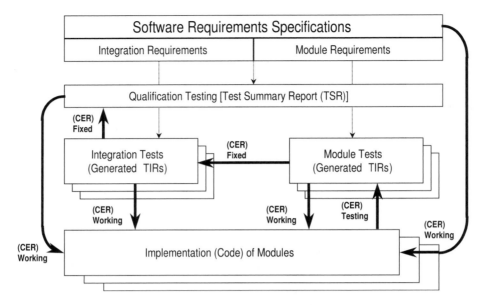

Figure 6-1. Software test progression.

Qualification Test. This section should describe qualification test. An example is:

All of the requirement test items (refer to the Software Requirements Traceability Matrix) are to be tested as qualification-level tests. For details of the procedures and setup, see the [PROJECT ABBREVIATION] Qualification Software Test Description. The resulting output of qualification test is the [PROJECT ABBREVIATION] Software Test Report (STR) and its attached Test Problem Report (TPR). If qualification test is passed and its results accepted by the customer the [PROJECT ABBREVIATION] software will be ready for beta release. Following Qualification Testing, the customer will review results. Signature of acceptance initiates product delivery and start of Installation Beta Test.

For the qualification-level tests, the following classes of tests will be used:
Check for correct handling of erroneous inputs
Check for maximum capacity
User interaction behavior consistency
Retrieving data
Saving data
Display screen and printing format consistency
Check interactions between modules
Measure time of reaction to user input
Functional flow

Integration Test. This section should describe the integration test. An example is:

All of the modules to be integration tested (refer to the Software Requirements Traceability Matrix) will be tested using integration-level test methodology. For details of the procedures and setup, see the [PROJECT ABBREVIATION] Integration Software Test Description. The resulting outputs of this test are Internal Test Reports (ITRs) or System Integration Test Reports (SITRs). When these integration tests are all passed, the [PROJECT ABBREVIATION] software will be ready for qualification-level testing.

For the integration-level tests, the following classes of tests will be used:
  User interaction behavior consistency
  Display screen and printing format consistency
  Check interactions between modules
  Measure time of reaction to user input

Module Test. This section should describe the module test. An example is:

All of the modules to be tested (refer to the Software Requirements Traceability Matrix) are to be tested using defined module-level test methodology. For details regarding test procedures and setup, see the [PROJECT ABBREVIATION] Module Software Test Description. The resulting outputs of this test are Internal Test Reports (ITRs) or Unit Test Reports (UTRs). When all of the module tests for a module are passed, the module is ready for integration-level testing.
  For the module-level tests, the following classes of tests will be used:
    Check for correct handling of erroneous inputs
    Check for maximum capacity
    User interaction behavior consistency
    Retrieving data
    Saving data
    Display screen and printing format consistency
    Measure time of reaction to user input

Installation Beta Test. This section should describe installation beta testing. An example is:

Following qualification testing, the customer will review results. Signature of acceptance initiates product delivery and start of Installation Beta Test. Identified tests (refer to the Software Requirements Traceability Matrix) will be tested using defined installation-level Beta Test procedures. Some of these tests maybe documented in the Product Packaging Information in Appendix C. This test methodology will be applied at each beta test site. For details regarding test procedures and setup, see the [PROJECT ABBREVIATION] Installation Software Test Description. Outputs of these tests are Test Problem Reports (TPRs). Following Installation Testing, the customer will review results. Signature of acceptance completes the [PROJECT ABBREVIATION] Version 1.0 project.
  For the installation-level tests, the following classes of tests will be used:
    User interaction behavior consistency
    Retrieving data
    Saving data
    Display screen and printing format consistency
    Check interactions between modules
    Measure time of reaction to user input
    Functional flow

Test Schedules. This section should include test milestones as identified in the software project schedule as well as any required deliverables associated with testing. Any additional milestones may be defined here as needed. If appropriate detail is provided in the Software Project Management Plan (SRMP), then this can be referenced. The following is as an example:

Test schedules are defined in the [PROJECT ABBREVIATION] Software Development Plan.

Risk Management. This section should identify all high-risk assumptions of the test plan. The contingency plans for each should be described and the management of risk items should be addressed.

Requirements Traceability. This section should provide information regarding the traceability of testing to the requirements and design of the software. A matrix, or database, is useful when meeting this traceability requirement. The following is an example:

> Refer to [PROJECT ABBREVIATION] Software Requirements Specification, Appendix A for information regarding requirements traceability.

Note. This section should provide any additional information not previously covered in the test development plan.

System Test Plan

The following information is based on IEEE Std 829-1998, IEEE Standard for Software Test Documentation [5]; IEEE Std 1012-1998, IEEE Standard for Software Verification and Validation [9]; IEEE Std 1028-2002, IEEE Standard for Software Reviews [12]; and IEEE Std 1044.1-1995, IEEE Guide to Classification of Software Anomalies [13], which have been adapted to support CMMI requirements. A document outline is provided in Table 6-15. Additional information is provided in the document template, *System Test Plan.doc,* which is located on the CD-ROM accompanying this book. Additional information in support of the work products commonly associated with this PA may be found in Appendix C.

The following provides section-by-section guidance in support of the creation of a system test plan. This guidance should be used to help define associated test requirements and should reflect the actual requirements of the implementing organization.

INTRODUCTION

This section should provide a brief introductory overview of the project and related testing activities described in the document.

Scope. This section contains two subsections.

Identification and Purpose. This subsection should provide information that uniquely identifies the system effort and the associated test plan. This can also be provided in the form of a unique test plan identifier. The following text provides an example:

> This system test plan is to detail the testing planned for the [Project Name] ([Project Abbreviation]) Version [xx], Statement of Work, [date], Task Order [to number], Contract No. [Contract #], and Amendments.
>
> The goal of [project acronym] development is to [goal]. This System will allow [purpose].
>
> Specifics regarding the implementation of these modules are identified in the [Project Abbreviation] System Requirements Specification (SysRS), with line item descriptions in the accompanying Requirements Traceability Matrix (RTM).

System Overview. This subsection should provide a summary of all system items and system features to be tested. The need for each item and its history may be addressed here

Table 6-15. System test plan document outline

Title Page
Revision Page
Table of Contents
1. Introduction
2. Scope
   2.1   Identification and Purpose
   2.2   System Overview
   2.3   Definitions, Acronyms, and Abbreviations
3. Referenced Documents
4. System Test Objectives
5. Kinds of System Testing
   5.1   Functional Testing
   5.2   Performance Testing
   5.3   Reliability Testing
   5.4   Configuration Testing
   5.5   Availability Testing
   5.6   Portability Testing
   5.7   Security and Safety Testing
   5.8   System Usability Testing
   5.9   Internationalization Testing
   5.10 Operations Manual Testing
   5.11 Load Testing
   5.12 Stress Testing
   5.13 Robustness Testing
   5.14 Contention Testing
6. System Test Environment
   6.1   Development Test and Evaluation
         6.1.1 System and Software Items
         6.1.2 Hardware and Firmware Items
         6.1.3 Other Materials
         6.1.4 Proprietary Nature, Acquirer's Rights, and Licensing
         6.1.5 Installation, Testing, and Control
         6.1.6 Participating Organizations
   6.2   Test Site(s)
7. Test Identification
   7.1   General Information
         7.1.1 Test Levels
         7.1.2 Test Classes
         7.1.3 General Test Conditions
         7.1.4 Test Progression
   7.2   Planned Testing
         7.2.1 Module Test
         7.2.2 Integration Test
         7.2.3 System Test
         7.2.4 Qualification Test
         7.2.5 Installation Beta Test
8. Test Schedules
9. Risk Management
10. Requirements Traceability
11. Notes
Appendix A. System Test Requirements Matrix
Appendix B. Module System Test Description
Appendix C. System Test Description
Appendix D. Qualification System Test Description
Appendix E. Installation System Test Description

as well. References to associated project documents should be cited here. The following provides an example:

> This document describes the System Test Plan (SysTP) for the [PROJECT ABBREVIATION] System. [PROJECT ABBREVIATION] documentation will include this SysTP, the [PROJECT ABBREVIATION] System Requirements Specification (SysRS), the System Requirements Traceability Matrix (RTM), the [PROJECT ABBREVIATION] Software Development Plan (SDP), the Software Design Document (SDD), the [PROJECT ABBREVIATION] User's Manual, the [PROJECT ABBREVIATION] System Administrator's Manual, and the [PROJECT ABBREVIATION] Data Dictionary.
>
> This SysTP describes the process to be used and the deliverables to be generated in the testing of the [PROJECT ABBREVIATION]. This plan, and the items defined herein, will comply with the procedures defined in the [PROJECT ABBREVIATION] Software Configuration Management (SCM) Plan and the [PROJECT ABBREVIATION] Software Quality Assurance (SQA) Plan. Any nonconformance to these plans will be documented as such in this SysTP.
>
> This document is based on the [PROJECT ABBREVIATION] System Test Plan template with tailoring appropriate to the processes associated with the creation of the [PROJECT ABBREVIATION]. The information contained in this SysTP has been created for [Customer Name] and is to be considered "For Official Use Only."
>
> Please refer to Section "6. Schedule" in the SDP for information regarding documentation releases.

Definitions, Acronyms, and Abbreviations. All relevant definitions, acronyms, and abbreviations should be included in this section.

Referenced Documents. This section should include all material referenced during the creation of the test plan.

System Test Objectives. Test objectives should include the validation of the application (i.e., to determine if it fulfills its system requirements specification), the identification of defects that are not efficiently identified during unit and integration testing, and the determination of the extent to which the system is ready for launch. This section should provide project status metrics (e.g., percentage of test scripts successfully tested). A description of some of the kinds of testing may be found in Appendix C, Software Process Work Products, in the Example of System Testing section.

System Test Environment. This section contains the following seven subsections.

Development Test and Evaluation. This subsection should provide a high-level summary of all key players, facilities required, and site of test performance. The following is provided as an example:

> System tests are to be performed at [company name], [site location]. All testing will be conducted in the development center [room number]. The following individuals must be in attendance: [provide list of performers and observers].

System and Software Items. This subsection should provide a complete list of all system and software items used in support of testing, including the versions. If the system is kept in an online repository, reference to the storage location may be cited instead of listing all items here. The following is an example:

System and software used in the testing of [PROJECT ABBREVIATION]: [System and software List]. For details of client and server System specifications see the [PROJECT ABBREVIATION] System Requirements Specifications (SysRS).

Hardware and Firmware Items. This subsection should provide a complete list of all hardware and firmware items used in support of testing. If this information is available in another project document, it may be referenced instead of listing all equipment. The following is an example:

[Hardware Items] are to be used during testing, connected together in a client server relationship through a TCP/IP-compatible network. For details of client and server hardware specifications see the [PROJECT ABBREVIATION] System Requirements Specifications (SysRS). For integration and qualification-level testing a representative hardware set as specified in the SysRS is to be used.

Other Materials. This subsection should list all other materials required to support testing.

Proprietary Nature, Acquirer's Rights, and Licensing. Any of the issues associated with the potential proprietary nature of the software, acquirer's rights, or licensing should be addressed in this subsection. An example is:

Licensing of commercial software is one purchased copy for each PC it is to be used on with the exception of [group software], which also requires licensing for the number of users/client systems. Some of the information in the data set is covered by the Privacy Act and will have to be protected in some manner.

Installation, Testing, and Control. This subsection should address environment requirements or actions required to support the installation of the application for testing. The following is provided as an example:

[Company Name] will acquire the software and hardware needed as per the [PROJECT ABBREVIATION] Software Development Plan (SDP) in order to run [PROJECT ABBREVIATION] in accordance with contract directions and limitations.

[Company Name] will install the supporting software and test it per the procedures in the [PROJECT ABBREVIATION] Installation System Test Description.

The [PROJECT ABBREVIATION] Change Enhancement Request (CER) tracking system will be used to determine eligibility for testing at any level. See the [PROJECT ABBREVIATION] Software Configuration Management Plan and [PROJECT ABBREVIATION] Software Quality Assurance Plan for details on CER forms and processes.

Participating organizations. This subsection should identify all groups responsible for testing. An example is:

The participating organizations are [Company Name], [customer name], and organizations at the beta test site.

These organizations may participate by completing the Test Plan Walk-through Checklist as described in Appendix C, Software Process Work Products.

Test Site(s). This section should provide a description of test sites. An example is:

The planned installation level Beta test sites are listed as follows: [Installation Beta Sites].

Test Levels. This section should describe the different levels of testing required in support of the development effort. See Appendix C, Test Design Specification, Test Case Specification, and Test Procedure Specification sections. The following is provided as an example:

> Tests are to be performed at the module, integration, installation, and qualification levels prior to release for beta testing. Please refer to test design specifications #xx through xx.

Test Classes. This section should provide a description of all test classes. A test case specification should be associated with each of these classes. A summary of the validation method, data to be recorded, data analysis activity, and assumptions, should be provided for each case. Example text is provided in Appendix C, Example Test Classes.

General Test Conditions. This section should describe the anticipated baseline test environment. An example is provided below:

> A sample real data set from an existing database is to be used during all tests. The sample real data set will be controlled under the configuration management system. A copy of the controlled data set is to be used in the performance of all testing.

Test Progression. This section should provide a description of the progression of testing as shown in Figure 6-2. A diagram is often helpful when attempting to describe the progression of testing. The information below is provided as an example:

> The Qualification Testing is a qualification-level test verifying that all requirements have been met. The module and integration tests are performed as part of the Implementation phase as elements and modules are completed. All module and integration tests must be

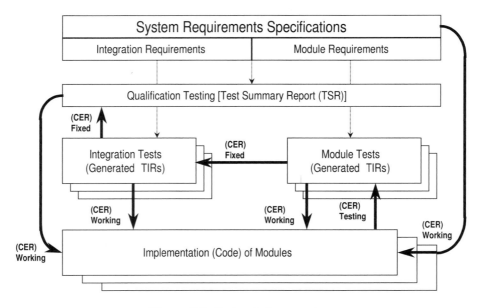

Figure 6-2 System test progression.

passed before performing System Testing. All module tests must be passed before performing associated integration level tests. The CER tracking system is used to determine eligibility for testing at a level. See the [PROJECT ABBREVIATION] Software Configuration Management Plan and [PROJECT ABBREVIATION] Software Quality Assurance Plan for details on CER forms and processes.

Planned Testing. This section should provide a detailed description of the type of testing to be employed. Examples text follows:

Module Test. All of the modules to be tested (refer to the Software Requirements Traceability Matrix) are to be tested using defined module-level test methodology. For details regarding test procedures and setup see the [PROJECT ABBREVIATION] Module Software Test Description. The resulting outputs of this test are Internal Test Reports (ITRs) or Unit Test Reports (UTRs). When all of the module tests are passed for a module, the module is ready for integration testing.

For the module-level tests, the following classes of tests will be used:
Check for correct handling of erroneous inputs
Check for maximum capacity
User interaction behavior consistency
Retrieving data
Saving data
Display screen and printing format consistency
Measure time of reaction to user input

Integration Test. All of the modules to be integration tested (refer to System Requirements Traceability Matrix) will be tested using integration-level test methodology. For details of the procedures and setup, see the [PROJECT ABBREVIATION] Integration System Test Description. The resulting outputs of this test are Internal Test Reports (ITRs) or System Integration Test Reports (SITRs).

When these integration tests are all passed, the [PROJECT ABBREVIATION] System will be ready for qualification level testing.

For the integration-level tests, the following classes of tests will be used:
User interaction behavior consistency
Display screen and printing format consistency
Check interactions between modules
Measure time of reaction to user input

System Test. All of the requirement test items (refer to the System Requirements Traceability Matrix) are to be tested as system-level tests. For details of the procedures and setup, see the [PROJECT ABBREVIATION] System Test Plan Description. The resulting output of system test is the [PROJECT ABBREVIATION] System Test Report (SysTR) and its attached Test Problem Report (TPR). If system test is passed and its results accepted by the [STAKEHOLDER] the [PROJECT ABBREVIATION] System will be ready for Qualification Testing. Following Qualification Testing, the customer will review the results. Signature of acceptance initiates product delivery and start of Installation Beta Test.

Qualification Test. All of the requirement test items (refer to the System Requirements Traceability Matrix) are to be tested as qualification-level tests. For details of the procedures and setup, see the [PROJECT ABBREVIATION] Qualification System Test Description. The resulting output of the qualification test is the [PROJECT ABBREVIATION] System Test Report (SysTR) and its attached Test Problem Report (TPR). If the qualification test is passed and its results accepted by the customer, the [PROJECT ABBREVIATION] System will be ready for beta release. Following Qualification Testing, the customer will review results. Signature of acceptance initiates product delivery and start of Installation Beta Test.

For the qualification-level tests the following classes of tests will be used:
Check for correct handling of erroneous inputs
Check for maximum capacity
User interaction behavior consistency
Retrieving data
Saving data
Display screen and printing format consistency
Check interactions between modules
Measure time of reaction to user input
Functional flow

Installation Beta Test. Following qualification testing, the customer will review results. Signature of acceptance initiates product delivery and start of Installation Beta Test. Identified tests (refer to the System Requirements Traceability Matrix) will be tested using defined installation-level Beta Test procedures. Some of these tests may be documented in the Product Packaging Information in Appendix C. This test methodology will be applied at each beta test site. For details regarding test procedures and setup see the [PROJECT ABBREVIATION] Installation System Test Description. Outputs of these tests are Test Problem Reports (TPRs). Following Installation Testing, the customer will review results. Signature of acceptance completes the [PROJECT ABBREVIATION] Version 1.0 project.

For the installation-level tests, the following classes of tests will be used:
User interaction behavior consistency
Retrieving data
Saving data
Display screen and printing format consistency
Check interactions between modules
Measure time of reaction to user input
Functional flow

Test Schedules. This section should include test milestones as identified in the system project schedule as well as any required deliverables associated with testing. Any additional milestones may be defined here as needed. If appropriate detail is provided in the Software Project Management Plan (SRMP), then this can be referenced. The following is provided as an example:

Test schedules are defined in the [PROJECT ABBREVIATION] Software Development Plan.

Risk Management. This section should identify all high-risk assumptions of the test plan. The contingency plans for each should be described and the management of risk items should be addressed.

Requirements Traceability. This section should provide information regarding the traceability of testing to the requirements and design of the system. A matrix, or database, is useful when meeting this traceability requirement. The following is provided as an example:

Refer to [PROJECT ABBREVIATION] System Requirements Specification, Appendix A, for information regarding requirements traceability.

Note. This section should provide any additional information not previously covered in the test development plan.

ORGANIZATIONAL PROCESS FOCUS

The purpose of the Organizational Process Focus (OPF) is to plan and implement organizational process improvement based on a thorough understanding of the current strengths and weaknesses of the organization's processes and process assets.

OPF is the first Process Area (PA) in the Process Management Category, which for Maturity Level 3 contains Organizational Process Definition and Organizational Training. OPF starts with Senior Management business objectives and provides resources and coordination to Organizational Process Definition. Organizational Process Definition provides standard processes and assets to Organizational Training and to the other process categories: Project Management, Engineering, and Support.

OPF establishes and maintains the set of standard processes and other assets based on the organization's process needs and objectives. Experiences and work products that are the result of performing these defined processes are incorporated into the organization's set of standard processes and other assets. The responsibility for facilitating and managing the organization's process-improvement activities is typically assigned to a software engineering process group, whose planning for process improvement results in a process-improvement plan. The process group can be an important part of the organization's approach to achieving Maturity Level 2.

The OPF PA results in several typical work product or artifacts, some of which are expected in other PAs. IEEE Standards 1058, IEEE Standard for Software Project Management Plans [15]; 730, IEEE Standard for Software Quality Assurance Plans [3]; and 1028, IEEE Standard for Software Reviews [12] provide detailed guidance in support of process plans and reviews. Additional information in support of the work products associated with this PA may be found in Appendix C, Organizational Process Improvement Checklist Using IDEAL, Organization Process Appraisal Checklist, and Process Lessons Learned sections.

CMMI®-SW Goals

SG1. Strengths, weaknesses, and improvement opportunities for the organization's processes are identified periodically and as needed.

Organizations are assessed or audited by using a standard (e.g., ISO 9001) or process model (e.g., CMMI) to determine the strengths and weaknesses at the process level. This information is prioritized by analysis of the organization's needs.

SG2. Improvements are planned and implemented, organizational process assets are deployed, and process-related experiences are incorporated into the organizational process assets.

An improvement process is selected, such as SEI's I.D.E.A.L. process [54], and changes are managed, just as in any development project, with the involvement of the process owners.

GG 3. The process is institutionalized as a defined process.

CMMI®-SW specifically requires that the OPF PA be institutionalized as a defined process, so the project teams have an audience to share their lessons learned and benefit from using improved processes and document templates from the organization's set of standard processes. These standard processes are tailored according to the organization's tailoring guidelines. Table 6-16 describes the specific and generic goals and practices, typical work products, and commonly associated plans or artifacts in support of the PA.

Engineering Process Group Charter

The following is provided as an example of the type of information typically provided in an Engineering Process Group (EPG) Charter document. The information provided here is for illustrative purposes; the EPG Charter of any organization should reflect the unique requirements of the developing organization.

Purpose.   This section should describe the rationale and scope of the EPG Charter. An example is:

> This charter formally empowers the Engineering Process Group (EPG) to manage the released content of corporate software engineering processes at [Company Name].

Vision/Mission Statement.   This section should describe the mission of the EPG, including major areas of responsibility. This is an example:

> The EPG provides management control of [Company Name] engineering processes. It:
>   a. Approves release into general use of engineering process assets
>   b. Manages proposed improvements to the engineering process assets as necessary
>   c. Provides tactical guidance on asset creation and maintenance priorities
>   d. Adjudicates controversial or unprecedented process tailoring decisions, as required

Authority and Responsibility.   This section should provide a description of EPG authority and responsibility. An example is:

> The EPG has authority to define all systems and software engineering process assets in each released set comprising the [Company Name] system and software engineering process. The EPG is responsible for the management and implementation of that asset set.

Membership and Organization.   A description of all voting and nonvoting members should be provided. The terms that each member is responsible for serving on the EPG should also be described. It is important to address what may be considered a quorum to convene meetings. An example of defining a quorum is:

> A quorum for the meeting shall consist of the Chair, SQA representative, and 60% of the voting membership.

Activities and Function.   All activities should be described in this part of the charter document. These activities and functions may include a description of process development, procedures for fast tracking processes, typical review and approval timelines, and supporting procedures.

Table 6-16. Organizational process focus goals and practices [61]

Specific and generic goals/practices and typical work products	Plan or artifact
SG1. Determine Process-Improvement Opportunities	
SP 1.1. Establish Organizational Process Needs	
Organization's process needs and objectives	Organizational process improvement plan, organizational process improvement checklist using IDEAL
SP 1.2. Appraise the Organization's Processes	
Plans for the organization's process appraisals	Organizational process improvement plan, appraisal plan, organization process appraisal checklist
Appraisal findings that address strengths and weaknesses of the organization's processes	Organizational process improvement plan
Improvement recommendations for the organization's processes	Organizational process improvement plan, process lessons learned
SP 1.3. Identify the Organization's Process Improvements	
Analysis of candidate process improvements	Organizational process improvement plan
Identification of improvements for the organization's processes	Organizational process improvement plan
SG2. Plan and Implement Process-Improvement Activities	
SP 2.1. Establish Process Action Plans	
Organization's approved process action plans	Organizational process improvement plan
SP 2.2. Implement Process Action Plans	
Commitments among the various process action teams	Organizational process improvement plan
Status and results of implementing process action plans	Organizational process improvement plan
Plans for pilots	
SP 2.3. Deploy Organizational Process Assets	
Plans for deploying the organizational process assets and changes to organizational process assets	Organizational process improvement plan
Training materials for deploying the organizational process assets and changes to organizational process assets	Organizational process improvement plan
Documentation of changes to the organizational process assets	Organizational process improvement plan
Support materials for deploying the organizational process assets and changes to organizational process assets	Organizational process improvement plan
SP 2.4. Incorporate Process-Related Experiences into the Organizational Process Assets	
Process-improvement proposals (PIP)	Organizational process improvement plan, PIP artifacts
Measurements on the organizational process assets	Organizational process improvement plan
Improvement recommendations for the organizational process assets	Organizational process improvement plan *(continued)*

Table 6-16. *Continued*

Specific and generic goals/practices and typical work products	Plan or artifact
SG2. Plan and Implement Process-Improvement Activities (*cont.*)	
SP 2.4. Incorporate Process-Related Experiences (*cont.*)	
Records of the organization's process-improvement activities	Organizational process improvement plan
Information on the organizational process assets and improvements to them	Organizational process improvement plan
GG3. Institutionalize a Defined Process	
GP 2.1. Establish an organizational policy	
Establishes organizational expectations for determining process-improvement opportunities for the processes being used and for planning and implementing process-improvement activities across the organization	Organizational policy
GP 2.2. Plan the process	
The plan for performing the organizational process focus process, which is often called "the process-improvement plan," differs from the process action plans described in specific practices in this process area	Organizational process improvement plan
GP 2.3. Provide resources	
Identification of resources used in support of organizational process focus activities	Organizational process improvement plan
GP 2.4. Assign responsibility	
Two groups are typically established and assigned responsibility for process improvement: (1) a management steering committee for process improvement to provide senior-management sponsorship; and (2) a process group to facilitate and manage the process-improvement activities	Organizational process improvement plan, EPG Charter
GP 2.5. Train people	
Project plan or training plan	Organizational process improvement plan
GP 2.6. Manage configurations	
Description of how work products and all associated artifacts in support of organizational process focus activities are placed under configuration management	Organizational process improvement plan
GP 2.7. Identify relevant stakeholders	
Describe activities and feedback mechanisms	Organizational process improvement plan, EPG Charter
GP 2.8. Monitor and control the process	
Description may be described in engineering process group charter	EPG Charter

Table 6-16. *Continued*

Specific and generic goals/practices and typical work products	Plan or artifact
GG3. Institutionalize a Defined Process (*cont.*)	
GP 2.9. Objectively evaluate adherence	
Individuals external to the organization or the engineering process group may conduct reviews	Organizational process improvement plan
GP 2.10. Review status with higher-level Management	
These reviews are typically in the form of a briefing presented to the management steering committee by the process group and the process action teams	Status review briefing
GP 3.1. Establish a defined process	
Establish and maintain a description of the organizational process focus process	Organizational process improvement plan
GP 3.2. Collect improvement information	Software measurement and metrics plan Organizational process improvement plan

Process Inventory and Characteristics

For assessments and evaluations, the SEI has published the Standard CMMI® Appraisal Method for Process Improvement (SCAMPI) and the Appraisal Requirements for CMMI® (ARC). CMMI® Assessments and evaluations allow organizations to:

1. Gain insight into their engineering capabilities through the identification of process strengths and weaknesses
2. Relate process strengths and weaknesses to the CMMI® model
3. Prioritize their improvement plans
4. Focus on improvements that are most beneficial
5. Derive capability and maturity level ratings
6. Identify risk relative to capability/maturity determinations [57]

ARC

The Appraisal Requirements for CMMI® (ARC) defines what is required for the development, definition, and use of appraisal methods in support of compliance with the CMMI® [60]. The ARC supports both CMMI® representations, staged and continuous, and supports assessments (for internal process improvement) and capability evaluations (for source selection and/or process monitoring). The ARC has formalized requirements for three classes of appraisal methods by mapping CMMI® requirements to each method. It provides organizations the freedom to develop an appraisal methodology that works best for their organization. Table 6-17 provides a summary of these three appraisal classes.

Class A describes a full appraisal, usually performed by a team of six to 10 people, primarily drawn from inside the organization being appraised. This appraisal method is expected to be the most accurate, maximize the buy-in from the appraisal participants, and

Table 6-17. Summary of the characteristics of the three CMMI® appraisal classes [60]

Characteristics	Class A	Class B	Class C
Usage Mode	• Rigorous and in-depth investigation of process(es) • Basis for improvement activities	• Initial (first-time) • Incremental (partial) • Self-assessment	• Quick look • Incremental
Advantages	• Thorough coverage • Strengths and weaknesses for each PA investigated • Robustness of method with consistent, repeatable results • Provides objective view	• Organization gains insight into own capability • Provides a starting point or focuses on areas that need most attention • Promotes buy-in	• Inexpensive • Short duration • Rapid feedback
Disadvantages	• Demands significant resources	• Does not emphasize depth of coverage and rigor and cannot be used for level rating	• Provides less buy-in and ownership of results • Not enough depth to fine-tune process-improve-ment plans
Sponsor	• Senior manager of organizational unit	• Any manager sponsoring an SPI program	• Any internal manager
Team Composition	• External and internal	• External or internal	• External or internal
Team Size	• 4–10 persons + assessment team leader	• 1–6 + assessment team leader	• 1–2 + assessment team leader
Team Qualifications	• Experienced	• Moderately experienced	• Moderately experienced
Assessment Team Leader Requirements	• Lead assessor	• Lead assessor or person experienced in method	• Person trained in method

will leave the organization with the best understanding of their weaknesses and any strengths that should be shared. The Standard CMMI® Appraisal Method for Process Improvement (SCAMPI) describes a Class A appraisal. [66]

SCAMPI

SCAMPI is designed to provide organizations with quality ratings relative to CMMI® models. It supports both internal process improvement assessments and external capability determinations, and satisfies all Class A appraisal ARC requirements. Assessment results can be used to support internal process improvement evaluation, supplier selection, and process monitoring [66]. Table 6-18 describes the SCAMPI modes of usage.

Table 6-18. SCAMPI modes of usage

Usage Mode	Description	Applications
Internal Process Improvement	Appraisals are used to evaluate internal processes: • Baseline capability/maturity levels • Establish or update a process-improvement program • Measure progress in program	• Measuring progress • Conducting audits • Appraising specific projects • Preparing for external appraisals
Supplier Selection	• Discriminator for supplier selection • Baseline for subsequent process monitoring	• Characterize process-related risk of supplier selection • Assist in supplier selection
Process Monitoring	• Help the sponsoring organization tailor its contract or process-monitoring efforts by allowing it to prioritize efforts based on the observed strengths and weaknesses of supplying orgs processes • Focus on long-term teaming relationship between sponsor and supplier organizations	• Serving as input for incentive/award fee • Risk management

As an ARC Class A method, SCAMPI can be used to generate ratings as benchmarks to compare maturity levels or capability levels across organizations. SCAMPI is an integrated appraisal method that can be applied in the context of internal process improvement, supplier selection, or process monitoring. SCAMPI offers a rigorous methodology capable of achieving high accuracy and reliability of appraisal results through the collection of objective evidence from multiple sources. It provides detailed guidance as to what should be measured and how the assessment should be carried out.

Process Action Plan (PAP)

Process action plans usually result from appraisals and document how specific improvements targeting the weaknesses uncovered by an appraisal will be implemented. In cases in which it is determined that the improvement described in the process action plan should be tested on a small group before deploying it across the organization, a pilot plan is generated. Finally, when the improvement is to be deployed, a deployment plan is used. This plan describes when and how the improvement will be deployed across the organization.

The following is provided as an example of the type of information typically provided in a Process Action Plan (PAP) document. The information provided here is for illustrative purposes; any specific PAP of an organization should reflect the unique requirements of the developing organization.

Identification. This section should provide summary information that may be used when referring to the document. This should include the author, the date, and any unique tracking number.

Instructions. This section should provide directions for using the PAP, including submission procedures.

Issue. This section should provide a description of the process area being defined or improved. If this is an improvement suggestion, identify the process being used in the context of the proposal being submitted. If a specific tool, procedure, or other EPG asset is being addressed, describe it here as well.

Process Action Team. This section should identify the personnel who will implement the actions.

Process Improvement. This section should provide the strategies, approaches, and actions to address the identified process improvements

Improvement. This section should describe the process or technological improvement being proposed to support resolution of the issue described above.

ORGANIZATIONAL PROCESS DEFINITION

The purpose of Organizational Process Definition (OPD) is to establish and maintain a usable set of organizational process assets and other assets based on the process needs and objectives of the organization. These other assets include descriptions of processes and process elements, descriptions of life-cycle models, process-tailoring guidelines, process-related documentation, and the Measurement Repository, which has its own definitions and data.

OPD is the second of three Process Areas (PAs) in support of the Process Management Category, for Maturity Level 3, and receives resources and coordination from Organization Process Focus. OPD provides standard processes and assets to Organizational Training and to the other process categories: Project Management, Engineering, and Support. Organizational Training develops or obtains training that develops the required skills.

The process asset library is a collection of items maintained by the organization for use by the people and projects of the organization. This collection of items includes descriptions of processes and process elements, descriptions of life-cycle models, process-tailoring guidelines, process-related documentation, and data.

The OPD PA results in several typical work product or artifacts, some which are expected in other PAs. IEEE Standards 1074, IEEE Standard for Developing Life Cycle Processes [19]; 12207.0, IEEE Standard for Information Technology—Software Life Cycle Processes [39]; and ISO/IEC 15288, Systems Engineering—System Life Cycle Processes [86] provide detailed guidance in support of process definitions, whereas IEEE Standards 982.1, IEEE Standard Dictionary of Measures to Produce Reliable Software [7]; 1061, IEEE Standard for a Software Quality Metrics Methodology [16]; and (pending adoption) ISO/IEC 14143.1, Information Technology—Software measurement—Functional Size Measurement—Definition of Concepts [86] provide detailed guidance in support of process-measurement definitions. Additional information in support of the work products associated with this PA are also included in Appendix C, Organizational Policy Process Definition Form, Process Asset Library Catalog, and Process versus Product Metrics sections.

IEEE Std 12207.0, Standard for Information Technology—Software Life Cycle Processes [39], describes 17 processes spanning the entire life cycle of a software product or service. Even if an organization's processes were defined using other sources, this standard is useful in characterizing the essential characteristics of these software processes and should be considered prior to the implementation of process improvement activities. Referencing this standard and reviewing what is required for each of these primary process areas can provide additional guidance in support of the activities associated with organizational process definition. It is important that IEEE 12207 be considered prior to the implementation of any process-improvement activities associated with organizational process definition.

In less mature organizations, both the Process Asset Library and Measurement Repository start as a simple spreadsheet inventory, maintained by the Project Manager.

It is important to clarify the relationship between IEEE Std 1074, Standard for Developing Life Cycle Processes, and the 12207 series of standards. The following is an excerpt from Annex I of IEEE/EIA 12207.0-1996:

> IEEE Std 1074 was primarily written for an organizational process architect; the individual responsible for establishing the software life cycle to be followed on a particular project or set of projects. The standards require that the process architect identify a set of available software life cycle models (e.g., prototyping, spiral, and waterfall). The process architect then selects one of these models to be used on the project. The activities of the standards are then mapped in time order onto the model. Assigning owners and schedule to the activities completes what the standard defines as the project software life cycle.
>
> IEEE 12207 establishes a common framework for the life cycle of software in terms of the processes that can be employed to (1) acquire, supply, develop, operate, and maintain software; (2) manage, control, and improve the processes; and (3) provide the basis for world trade in software. 12207 places requirements upon the characteristics of a designated set of life cycle processes, but does not specify the detailed implementation of those processes; a process architect may find IEEE Std 1074 to be useful in developing organizational processes complying with the requirements of IEEE Std 12207. [39]

CMMI®-SW Goals

SG 1. Establish Organizational Process Assets

The organization must identify, collect, and/or develop process assets for use by all projects. These assets may have multiple versions to meet the needs of different application domains, life-cycle models, methodologies, and tools. One key asset is the Life-Cycle Model Description. Another asset describes how projects can tailor these assets to meet project objectives. Tailoring guidelines describe mandatory and optional process elements that provide the balance of flexibility and standardization to ensure appropriate consistency in the processes across the organization. Another key asset is the Organization's Measurement Repository, which is used to collect both product and process measures. Both the Measurement Repository and Process Asset Library must be established and maintained.

GG 3. The process is institutionalized as a defined process.

CMMI®-SW specifically requires that the Requirements Development PA be institutionalized as a defined process, so the project team can tailor their processes and document templates from the organization's set of standard processes according to the organization's tailoring guidelines.

GP 3.1. Establish a Defined Process

The closest related CMMI Generic Practice is GP 3.1—Establish a Defined Process. This ensures that OPD provides a defined process for all CMMI Process Areas. Table 6-19 de-

Table 6-19. Organizational process definition goals and practices [61]

Specific and generic goals/practices and typical work products	Plan or artifact
SG1. Establish Organizational Process Assets	
SP 1.1. Establish Standard Processes	
Organization's set of standard processes	Organizational process improvement plan, Organizational policies, process-definition form
SP 1.2. Establish Life-Cycle Model Descriptions	
Descriptions of life-cycle models	Organizational process improvement plan
SP 1.3. Establish Tailoring Criteria and Guidelines	
Tailoring guidelines for the organization's set of standard processes	Tailoring guidelines
SP 1.4. Establish the Organization's Measurement Repository	
Definition of the common set of product and process measures for the organization's set of standard processes	Software measurement and metrics plan, process versus product metrics
Design of the organization's measurement repository	Software measurement and metrics plan
Organization's measurement repository	Software measurement and metrics plan
Organization's measurement data	Measurement data
SP 1.5. Establish the Organization's Process Asset Library	
Design of the organization's process asset library	Organizational process improvement plan
Organization's process asset library	Organizational process improvement plan, organization's process asset library catalog
Selected items to be included in the organization's process asset library	Organizational process improvement plan
Catalog of items in the organization's process asset library	Organizational process improvement plan, process asset library catalog
GG3. Institutionalize a Defined Process	
GP 2.1. Establish an organizational policy	
Establishes organizational expectations for establishing and maintaining a set of standard processes for use by the organization and making organizational process assets available across the organization	Organizational policy
GP 2.2. Plan the process	
Typically, this plan for performing the organizational process definition process is a part of the organization's process-improvement plan	Organizational process improvement plan
GP 2.3. Provide resources	
Identification of resources used in support of organizational process definition activities	EPG Charter

Table 6-19. *Continued*

Specific and generic goals/practices and typical work products	Plan or artifact
GG3. Institutionalize a Defined Process (*cont.*)	
GP 2.4. Assign responsibility	
Two groups are typically established and assigned responsibility for process improvement: (1) a management steering committee for process improvement that provides senior-management sponsorship; and (2) a process group to facilitate and manage the process-improvement activities	EPG Charter
GP 2.5. Train people	Software project management plan or training plan
GP 2.6. Manage configurations	
Description of how work products and all associated artifacts in support of organizational process definition activities are placed under configuration management	EPG Charter
GP 2.7. Identify relevant stakeholders	
Describe activities and feedback mechanisms	EPG Charter
GP 2.8. Monitor and control the process	
Description may be described in engineering process group charter	EPG Charter
GP 2.9. Objectively evaluate adherence	
Individuals external to the organization or the engineering process group may conduct reviews	Organizational process improvement plan
GP 2.10. Review status with higher-level management	
These reviews are typically in the form of a briefing presented to the management steering committee by the process group and the process action teams	Status review briefing
GP 3.1. Establish a defined process	
Establish and maintain a description of the organizational process definition process	Organizational process improvement plan
GP 3.2. Collect improvement information	Software measurement and metrics plan, or software project management plan

scribes the specific and generic goals and practices, typical work products, and commonly associated plans or artifacts in support of the PA.

Organization's Set of Standard Processes

The information provided here is based upon IEEE Std1074-1997, IEEE Standard for Developing Software Life Cycle Processes [19] and IEEE Std 12207.0, Standard for Information Technology—Software Life Cycle Processes [39]. Table 6-20 describes the three categories of life cycle processes that are found in IEEE 12207.0.

This three-dimensional view of standard processes is supported by Figure 6-3. This figure is from IEEE 12207 and also provides an illustration of representative viewpoints.

Tailoring Guidelines

Processes should be tailored for a project, as no two projects are the same. Tailoring is the adaptation of an organization's set of standard processes for use on a specific project by using tailoring guidelines defined by an organization. The purpose of tailoring the organizational process asset for a specific project is to ensure that the appropriate amount of effort is devoted to the appropriate activities so as to reduce project risk to an acceptable level while at the same time making most cost-effective use of engineering talent.

IEEE 12207.0, Annex A, Tailoring Process has four major steps:

1. Identify project environment—strategy, activity, requirements
2. Solicit inputs—from users, support team, potential bidders
3. Select processes, activities, documentation, responsibilities
4. Document tailoring decisions and rationale

Elaborations of these general steps in tailoring the organizational process asset for a project are:

1. Characterize the project environment:
   - Size of the development team
   - Strategy
   - Critical factors

Table 6-20. IEEE 12207 Life Cycle Processes [39]

Primary life cycle processes	Acquisition process
	Supply process
	Development process
	Operation process
	Maintenance process
Supporting life cycle processes	Documentation process
	Configuration management process
	Quality assurance process
	Verification process
	Validation process
	Joint review process
	Audit process
	Problem resolution process
Organizational life cycle processes	Management process
	Infrastructure process
	Improvement process
	Training process

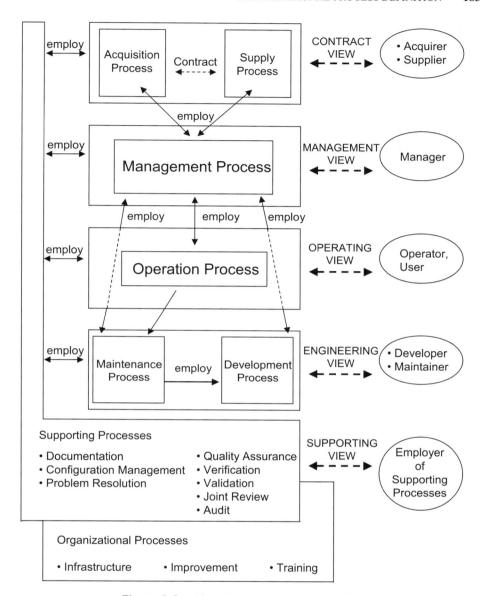

Figure 6-3. Life cycle processes and views [39].

- Project/customer requirements
- Life cycle strategy: waterfall, evolutionary, spiral, and so on
- Life cycle activity: prototyping, maintenance
- Software characteristics: COTS, reuse, embedded firmware
- Project policies, languages, hardware reserve, culture
- Acquisition strategy: contract type, contractor involvement

2. Determine the cost targets and risk tolerance of the project.

3. Identify the organization process asset from which tailoring is being considered.

4. For each committed deliverable asset, use best practices and professional judgment to identify the *form* the asset should take, and *level of detail* necessary to achieve the purpose of each organization process asset.

5. Assess whether any organization process assets, or their forms or their level of detail, are unaffordable given the project goals, cost targets, and level of tolerable risk.

6. Identify the level of detail needed for each process activity.

7. Document the tailoring planned for the organization process asset, and obtain approval. Typically, it is unwise to tailor the asset intent or objectives. What normally is tailored are: number of phases/activities, roles, responsibilities, document formats, and formality/frequency of reports or reviews.

8. Document the planned processes, products, and reviews. Describe the completion criteria for each process.

9. After project completion, provide example implementations, lessons learned, and improvements back to the organization to continually improve the organizational process assets.

ORGANIZATIONAL TRAINING

The purpose of Organizational Training (OT) is to develop the skills and knowledge of all project team members so they can perform their roles effectively and efficiently. OT is the third of three Process Areas (PAs) in the Process Management Category required for Maturity Level 3 and receives resources and coordination from Organizational Process Focus and standard processes and assets from Organizational Process Definition in order to develop or obtain training that develops the required skills.

OT identifies the organization's strategic and tactical training needs. These are common across projects and support groups. The main training components include a managed training-development program, documented plans, personnel with appropriate knowledge, and mechanisms for measuring the effectiveness of the training program. Additional information in support of the work products associated with this PA are also included in Appendix C, Training Request Form and Training Log sections.

An immature organization usually defines needed expertise based upon job positions.

CMMI®-SW Goals

SG 1. A training capability that supports the organization's management and technical roles is established and maintained.

The organization should establish the strategic training needs that are common across the projects. Based on the needs, the organization should establish an OT Tactical Plan to develop the training capability.

SG 2. Training necessary for individuals to perform their roles effectively is provided.

Using the training capability, the organization should deliver training according to its records and later assess the training effectiveness.

GG 3. The process is institutionalized as a defined process.

CMMI®-SW specifically requires that the OT PA be institutionalized as a defined process so the project team can derive their project training from the organization's set of standard processes according to the organization's tailoring guidelines.

GP 2.5. Train the people performing or supporting the process as needed.

The closest related CMMI® Generic Practice is GP 2.5—Train People, which ensures that training support is provided for all CMMI® Level 2 and 3 Process Areas. Table 6-21 describes the specific and generic goals and practices, typical work products, and commonly associated plans or artifacts in support of the PA.

Each of the CMMI-SW® Level 2 and Level 3 process areas has associated training requirements. The Level 2 and 3 Generic Practice 2.5 requires that individuals be trained appropriately. This Generic Practice is applied to all process areas and the organizational training plan should address each of these process areas.

Training Plan

This document template uses IEEE Std-12207.0, Standard for Information Technology Software Lifecycle Processes; IEEE Std- 12207.1, Standard for Information Technology Software Lifecycle Processes—Life Cycle Data; IEEE Std-12207.2, Standard for Information Technology Software Lifecycle Processes—Implementation Considerations [39]; and the U.S. DOE Training Plan Template* as primary reference material. Table 6-22 provides a suggested training plan document outline. Additional information is provided in Appendix C in support of this PA in the form of an example training request form and training log.

The following provides section-by-section guidance in support of the creation of a training plan. Ensuring that all individuals are properly trained is integral to the success of any software development effort. Additional information is provided in the document template, *Training Plan.doc,* which is located on the CD-ROM accompanying this book. This training plan template is designed to facilitate the development of a planned approach to organizational or project-related training.

Introduction. This section should describe the purpose of the training plan and the organization of the document. As described by IEEE Std 12207.2,

> The Training Process is a process for providing and maintaining trained personnel. The acquisition, supply, development, operation, or maintenance of software products is largely dependent upon knowledgeable and skilled personnel. For example: developer personnel should have essential training in software management and software engineering. It is, therefore, imperative that personnel training be planned and implemented early so that trained personnel are available as the software product is acquired, supplied, developed, operated, or maintained. [39]

Scope. This section should describe the scope and purpose of the training, such as initial training for system users, refresher training for the system maintenance staff, training for system administrators, and so on.

---

*U.S. Department of Energy, Training Plan Template, http://cio.doe.gov/ITReform/sqse/download/traintem. doc.

Table 6-21. Organizational training goals and practices [61]

Specific and generic goals/practices and typical work products	Plan or artifact
SG1. Establish an Organizational Training Capability	
SP 1.1. Establish the Strategic Training Needs	
Training needs	Training survey and results, training request form
Assessment analysis	Training survey and results
SP 1.2. Determine which Training Needs Are the Responsibility of the Organization	
Common project and support group training needs	Organizational training tactical plan
Training commitments	Organizational training tactical plan
SP 1.3. Establish an Organizational Training Tactical Plan	Organizational training tactical plan
SP 1.4. Establish Training Capability	Training materials and supporting artifacts
SG2. Provide Necessary Training	
SP 2.1. Deliver training	
Delivered training course	Records of delivered courses
SP 2.2. Establish training records	
Training records	Training records, training log
Training updates to the organizational repository	Organizational repository
SP 2.3. Assess training effectiveness	
Training effectiveness surveys	Organizational training plan
Training program performance assessments	Organizational training plan
Instructor evaluation forms	Organizational training plan
Training examinations	Organizational training plan
GG3. Institutionalize a Defined Process	
GP 2.1. Establish an organizational policy	
This policy establishes organizational expectations for identifying the strategic training needs of the organization and providing that training	Organizational training policy
GP 2.2. Plan the process	
The plan called for in this generic practice would address the comprehensive planning for all of the specific practices in this process area. In contrast, the organizational training tactical plan called for in the specific practice would address the periodic planning for the delivery of individual training offerings.	Organizational training procedures
GP 2.3. Provide resources	
Identification of resources used in support of organizational process training activities	Organizational training plan
GP 2.4. Assign responsibility	
Identify individual(s) responsible for organizational process training	Organizational training plan

Table 6-21. *Continued*

Specific and generic goals/practices and typical work products	Plan or artifact
GG3. Institutionalize a Defined Process (*cont.*)	
GP 2.5. *Train people*	Organizational training plan
GP 2.6. *Manage configurations*	
Description of how work products and all associated artifacts in support of organizational process training activities are placed under configuration management	Organizational training plan
GP 2.7. *Identify relevant stakeholders*	
Describe activities and feedback mechanisms	Organizational training plan
GP 2.8. *Monitor and control the process*	
Description may be provided in engineering process group charter	EPG Charter
GP 2.9. *Objectively evaluate adherence*	
Individuals external to the organization or the engineering process group may conduct reviews	Organizational training plan
GP 2.10. *Review status with higher-level management*	
These reviews are typically in the form of a briefing presented to the management steering committee by the process group and the process action teams.	Status review briefing
GP 3.1. *Establish a defined process*	
Establish and maintain a description of the organizational process training process	Organizational training plan
GP 3.2. *Collect improvement information*	Software measurement and metrics plan, or software project management plan

Objectives. This section should describe the objectives or anticipated benefits from the training. It is best to express all objectives as actions that the users will be expected to perform once they have been trained.

Background. This section should provide a general description of the project, and an overview of the training requirements.

References. This section should identify sources of information used to develop this document, including reference to the existing organizational training policy. Reference to the organizational training plan, if the plan being developed is tactical, should be provided as well. Any existing engineering process group (EPG) charter should be referenced to provide insight into process oversight activities if needed.

Definitions and Acronyms. This section should provide a description of all definitions and acronyms used within this document that may not be commonly understood or are unique in their application within this document.

Table 6-22. Training plan document outline

Title Page
Revision Page
Table of Contents
1.0 Introduction
     1.1 Scope
     1.2 Objectives
     1.3 Background
     1.4 References
     1.5 Definitions and Acronyms
2.0 Training Requirements
     2.1 Roles and Responsibilities
     2.2 Training Evaluation
3.0 Training Strategy
     3.1 Training Sources
     3.2 Dependencies/Constraints/Limitations
4.0 Training Resources
     4.1 Vendor Selection
     4.2 Course Development
5.0 Training Environment
6.0 Training Materials

Training Requirements. This section should describe the general work environment (including equipment), and the skills for which training is required (management, business, technology, etc.). The training audience should also be identified (category of user: upper management, system administrator, administrative assistant, etc.). It may also identify individuals or positions needing specific training. Include the time frame in which training must be accomplished. Identify whether the training requirements are common, cross-project, requirements or if the requirements are required in support of a unique project requirement.

Roles and Responsibilities. This section should identify the roles and responsibilities of the training staff and identify the individuals responsible for the management of the training development and implementation. It may also include the identification of other groups who may serve as consultants, such as members of the development team, experienced users, and so on. The individuals responsible for keeping training records and for updating any associated organizational training repository should be identified. If this information is provided in the associated project plan, a reference to this plan may be provided here.

Training Evaluation. The effectiveness of training must continually be evaluated. This section should describe how training evaluation will be performed. Evaluation tools, surveys, forms, and so on, should be included. Also describe the revision process with regard to the modification of the course and course materials resulting from the evaluations.

Training Strategy. This section should describe the type of training (e.g., classroom, CBT) and the training schedule (duration, sites, and dates). Some factors may include adequacy of training facilities, accommodations, need to install system files, modem/communication issues, physical access to buildings, escorts needed within facilities, and so on. It is suggested that a training log be developed to document and track information associated with individuals receiving training.

Training Sources. This section should identify the source or provider for the training. Training may be internal (course developed in-house) or external (contracted to external training agencies).

Dependencies/Constraints/Limitations. This section should identify all known dependencies constraints and/or limitations that could potentially affect training on the project.

Training Resources. This section should include hardware/software, instructor availability, training time estimates, projected level of effort, system documentation, and other resources required to familiarize the trainer with the system, produce training materials, and provide the actual training. The identification and availability of other resource groups should also be included.

Vendor Selection. This section should describe the criteria for training vendor selection.

Course Development. This section should describe the requirements for internal course development. This should include all related lifecycle activities associated with the development of the training.

Training Environment. This section should describe the equipment and conditions required for the training, including installations, facilities, and special databases (typically should be a separate, independent development/production environment). Also identify any actions required by other groups, such as trainees, to request training.

Training Materials. This section should describe the types of training materials required for the training. The training materials developed may include visuals for overhead projectors, handouts, workbooks, manuals, computerized displays, and demonstrations. The training materials and curriculum should accurately reflect the training objective.

Update/Revise Training Plan. Once the training plan is developed, it must be subjected to the same kind of configuration management process as the other system documentation. Training materials should keep up with system enhancements. Describe the change release process with regard to all training documentation.

INTEGRATED PROJECT MANAGEMENT

The purpose of Integrated Project Management (IPM) is to establish and manage the project and the involvement of the relevant stakeholders according to an integrated and defined process that is tailored from the organization's set of standard processes. IPM is part of the advanced Project Management Process Areas for Maturity Level 3 and interfaces with Project Planning, Project Monitoring and Control, and Supplier Agreement Management. IPM also provides risk inputs into Requirements Management. IPM provides Organizational Process Focus with lessons learned and improved assets.

IPM goes beyond the Project Planning purpose of establishing and maintaining plans that define project activities and goes beyond the Project Monitoring and Control purpose of providing an understanding of the project's progress so that appropriate corrective actions can be taken when the project's performance deviates significantly from the plan.

IPM establishes and maintains the project's defined process that is tailored from the organization's set of standard processes. IPM ensures that the project is managed using the project's defined process; the project uses and contributes to the organization's process assets. IPM takes the Organizational Process Assets and Tailoring guidelines from the Organizational Process Definition organization's Process Asset Library to develop the Project's Defined Processes and uses the organization's Process Definition measurement repository to develop estimates. IPM provides for the management of stakeholder involvement, along with identification, negotiation, and tracking of project critical dependencies, and the resolution of project coordination issues.

IPM PA results in several typical work products or artifacts, some which are expected in other PAs. IEEE Standard 1490-2003, the IEEE Adoption of the Project Management Body of Knowledge [34], provides detailed guidance in support of the work breakdown structure (WBS), which groups project components that organize and define the total scope of the project. Additional information in support of the work products associated with this PA are also included in Appendix C, Status Reviews and Critical Dependencies Tracking sections.

In an immature organization, the project manager usually relies on other managers to knit together all commitments and consequences.

CMMI®-SW Goals

SG 1. The project is conducted using a defined process that is tailored from the organization's set of standard processes.

The integrated project team establishes and maintains the project's defined process. In addition, the project uses the organizational process assets and measurement repository for estimating and planning the project's activities that are integrated with other plans that affect the project. The project's defined processes manage the project using the project plan and the other plans that affect the project. As the project ends, its work products, measures, and documented experiences are offered to the organization's Process Assets Library and Measurement Repository.

SG 2. Coordination and collaboration of the project with relevant stakeholders is conducted.

The integrated project team manages the involvement of the relevant stakeholders in the project, including their participation to identify, negotiate, and track critical dependencies and resolve issues.

GG 3. The process is institutionalized as a defined process

CMMI®-SW specifically requires that the IPM PA be institutionalized as a defined process, so the project team can tailor their processes and document templates from the organization's set of standard processes according to the organization's tailoring guidelines. Table 6-23 describes the specific and generic goals and practices, typical work products, and commonly associated plans or artifacts in support of the PA.

Table 6-23. Integrated project management goals and practices

Specific and generic goals/practices and typical work products	Plan or artifact
SG1. Use the Project's Defined Process	
SP 1.1. Establish the Project's Defined Process	
The project's defined process	Software project management plan
SP 1.2. Use Organizational Process Assets for Planning Project Activities	
Project estimates and plans	Software project management plan, WBS
SP 1.3. Integrate Plans	
Integrated plans	Software project management plan and all supporting planning documentation, integrated plan's work-breakdown structure
SP 1.4. Manage the Project Using the Integrated Plans	
Work products created by performing the project's defined process	Software project management plan, all supporting work products, critical dependencies tracking
Collected measures and progress records or reports	Software project management plan or software measurement and metrics plan
Revised requirements, plans, and commitments	Requirements management plan, requirements specification, requirements tracking database
Integrated plans	Software project management plan and all supporting planning documentation
SP 1.5. Contribute to the Organizational Process Assets	
Proposed improvements to the organizational process assets	Organizational policy
Actual process and product measures collected from the project	Software project management plan or software measurement and metrics plan
Documentation	Tailored processes/procedures
SG2. Coordinate and Collaborate with Relevant Stakeholders	
SP 2.1. Manage Stakeholder Involvement	
Agendas and schedules for collaborative activities	Software project management plan, status reviews
Documented issues	Software project management plan
Recommendations for resolving relevant stakeholder issues	Software project management plan
SP 2.2. Manage Dependencies	
Defects, issues, and action items resulting from reviews with relevant stakeholders	Software project management plan
Critical dependencies	Software project management plan, critical dependencies tracking
Commitments to address critical dependencies	Software project management plan, critical dependencies tracking

(*continued*)

Table 6-23. *Continued*

Specific and generic goals/practices and typical work products	Plan or artifact
SG2. Coordinate and Collaborate with Relevant Stakeholders (*cont.*) *SP 2.2. Manage Dependencies (cont.)* Status of critical dependencies	Software project management plan, critical dependencies tracking, status reviews
SP 2.3. Resolve Coordination Issues Relevant stakeholder coordination issues Status of relevant stakeholder coordination issues	Software project management plan Software project management plan
GG3. Institutionalize a Defined Process *GP 2.1. Establish an organizational policy* This policy establishes organizational expectations for using the project's defined process and coordinating and collaborating with relevant stakeholders.	Organizational policy
GP 2.2. Plan the process Typically, this plan for performing the integrated project management process is a part of the project plan as described in the Project Planning process area.	Software project management plan
GP 2.3. Provide resources Identification of resources used in support of integrated project management activities	Software project management plan
GP 2.4. Assign responsibility Identify individual(s) responsible for integrated project management	Software project management plan
GP 2.5. Train people Train individuals responsible for integration of project management activities	Training plan or software project management plan
GP 2.6. Manage configurations	Software project management plan and configuration management plan
GP 2.7. Identify relevant stakeholders Describe activities and feedback mechanisms	Software project management plan
GP 2.8. Monitor and control the process Monitor and control, reporting all tailoring and changes associated with integrated project management	Software project management plan
GP 2.9. Objectively evaluate adherence Individuals external to the organization or the engineering process group may conduct reviews	SQA plan or software project management plan
GP 2.10. Review status with higher-level management These reviews are typically in the form of a briefing presented to the management steering committee by the process group and the process action teams.	Status review briefing

Table 6-23. *Continued*

Specific and generic goals/practices and typical work products	Plan or artifact

GG3. Institutionalize a Defined Process (*cont.*)
  GP 3.1. Establish a defined process
  This generic practice establishes and maintains a defined integrated project management process. The Establish the Project's Defined Process specific practice defines the project's defined process, which includes all processes that affect the project.
	Software project management plan

GP 3.2. Collect improvement information	Software measurement and metrics plan, or software project management plan

Work-Breakdown Structure (WBS)

A Work-Breakdown Structure (WBS) defines and breaks down all work associated with a project into manageable parts. It describes all activities that have to occur to accomplish the project. The WBS can serve as the foundation for the integration of project component schedules, budget, and resource requirements. IEEE Std 1490-2003, IEEE Guide—Adoption of the PMI Standard, A guide to the Project Management Body of Knowledge [34] recommends the use of the structure shown in Figure 6-4. Additional information in support of associated IPM work products is provided in Appendix C, Software Process Work Products.

RISK MANAGEMENT

The purpose of Risk Management (RSKM) is to identify potential problems before they occur, so that risk-handling activities may be planned and invoked as needed across the life of the product or project to mitigate adverse impacts on achieving objectives.

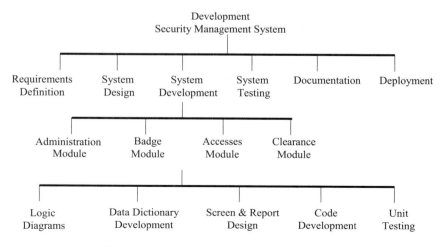

Figure 6-4. Example work-breakdown structure.

RSKM is part of the advanced Project Management Process Areas for Maturity Level 3 and interfaces with Project Planning, Project Monitoring and Control, and Supplier Agreement Management. RSKM goes beyond the Project Planning and Project Monitoring and Control risk identification and monitoring by taking a more continuing, forward-looking approach to managing risks with activities that include identification of risk parameters, risk assessments, and risk handling. RSKM is an evolution of Project Planning and Project Monitoring and Control specific practices to systematically plan, anticipate, and mitigate risks to proactively minimize their impact on the project. RSKM addresses issues that could endanger achievement of critical objectives. RSKM also interfaces with Requirements Development and Product Integration on technical solutions and technical external interfaces, where RSKM must consider both internal and external sources for cost, schedule, and technical risk. Additional information in support of the work products associated with this PA are also included in Appendix C: Risk Taxonomy Questionnaire, Risk Action Request, Risk Mitigation Plan, Risk Matrix, Description of Severity Levels, and Description of Probability Levels sections.

In an immature organization, RSKM practices occur in Project Planning, Project Monitoring and Control, Supplier Agreement Management, Technical Solution, and Product Integration process areas without overall planning or coordination. IEEE Std 12207.0, Standard for Information Technology—Software Life Cycle Processes [39], describes 17 processes spanning the entire life cycle of a software product or service. Even if an organization's processes were defined using other sources, this standard is useful in characterizing the essential characteristics of these software processes and should be considered prior to the implementation of process improvement activities. Referencing this standard and reviewing what is required for each of these primary process areas can provide additional guidance in support of the activities associated with risk management. It is important that IEEE 12207 be considered prior to the implementation of any process improvement activities associated with risk management activities.

CMMI®-SW Goals

SG 1. Prepare for Risk Management

The organization should prepare for risk management by determining risk sources and categories, along with risk parameters to analyze and categorize risks, and to control the risk management effort. The final preparation is the risk management strategy.

SG 2. Identify and Analyze Risks

The organization should identify, analyze, and document risks to determine their relative importance. Using the defined risk categories and parameters, evaluate and categorize each identified risk along and determine its relative priority.

SG3. Mitigate Risks

The organization should develop a risk mitigation plan and follow it up by periodically monitoring the status of each risk and, as appropriate, implementing the risk mitigation plan as defined by the risk management strategy.

GG 3. The process is institutionalized as a defined process.

CMMI®-SW specifically requires that the RSKM PA be institutionalized as a defined process, so the project team can tailor their processes and document templates from the organization's set of standard processes according to the organization's tailoring guidelines. Table 6-24 describes the specific and generic goals and practices, typical work products, and commonly associated plans or artifacts in support of the PA.

Risk Management Plan

Risk management focuses on the assessment and control (mitigation) of associated project risk. Risk assessment essentially involves the identification of all of the potential dangers, and their severity, that will affect the project and the evaluation of the probability of occurrence and potential loss associated with each item listed. The control of risk requires risk mitigation through the development of techniques and strategies for dealing with the highest-ordered risks. Risk control also involves the evaluation of the effectiveness of the techniques and strategies employed throughout the project life cycle. Table 6-25 provides a suggested outline for a risk management plan.

The following provides section-by-section guidance in support of the creation of a risk management plan. Ensuring that all risks are identified and controlled is a required CMMI® key process. Additional information is provided in the document template, *Risk Management Plan.doc,* which is located on the CD-ROM accompanying this book. This risk management plan template is designed to facilitate the development of a planned approach to risk management activities. This document template uses IEEE Std 1540-2001, IEEE Standard for Software Life Cycle Processes— Risk Management [37] and IEEE Std 12207.0 and 12207.1, Standards for Information Technology—Software life Cycle Processes [39] as primary reference material that has been adapted to support CMMI® requirements. Additional information in support of RM associated work products is provided in Appendix C.

Introduction. This section should provide the reader with a basic description of the project, including associated contract information and any standards used in the creation of the document. A description of the document approval authority and process should be provided.

Purpose. This section should provide a description of the motivation for plan implementation and the expectations for anticipated benefit. Define the perceived benefit from plan execution.

Scope. This section should describe the specific products and the portion of the software life cycle covered by the risk management plan. An overview of the major section of the risk management plan should also be provided.

Definitions and Acronyms. This section should provide a complete list of all acronyms and all relevant key terms used in the document that are critical to reader comprehension.

References. This section should list all references used in the preparation of this document.

Table 6-24. Risk management goals and practices [61]

Specific and generic goals/practices and typical work products	Plan or artifact
SG1. Prepare for Risk Management	
SP 1.1. Determine Risk Sources and Categories	
Risk source and categories list	Software project management plan, risk management plan, or risk taxonomy questionnaire
SP 1.2. Define Risk Parameters	
Risk evaluation, categorization, and prioritization criteria	Software project management plan, risk management plan, or risk taxonomy questionnaire
Risk management requirements	Software project management plan or risk management plan
SP 1.3. Establish a Risk Management Strategy	
Project risk management strategy	Software project management plan, risk management plan, or risk matrix and description of severity and probability levels
SG2. Identify and Analyze Risks	
SP 2.1. Identify Risks	
List of identified risks, including the context, conditions, and consequences of risk occurrence	Software project management plan or risk management plan
SP 2.2. Evaluate, Categorize, and Prioritize Risks	Software project management plan or risk management plan
SG3. Mitigate Risks	
SP 3.1. Develop Risk Mitigation Plans	
Documented handling options for each identified risk	Software project management plan, risk management plan, or risk action request
Risk mitigation plans	Software project management plan, risk management plan, or risk mitigation plan
Contingency plans	Software project management plan or risk management plan
List of those responsible for tracking and addressing each risk	Software project management plan, risk management plan, or risk action request
SP 3.2. Implement Risk Mitigation Plans	
Updated lists of risk status	Software project management plan or risk management plan
Updated assessments of risk likelihood, consequence, and thresholds	Software project management plan or risk management plan
Updated lists of risk handling options	Software project management plan or risk management plan
Updated list of actions taken to handle risks	Software project management plan or risk management plan
Risk mitigation plans	Software project management plan, risk management plan, or risk mitigation plan

Table 6-24. Risk management goals and practices [61]

Specific and generic goals/practices and typical work products	Plan or artifact
GG3. Institutionalize a Defined Process	
GP 2.1. Establish an organizational policy Establishes organizational expectations for defining a risk management strategy and identifying, analyzing, and mitigating risks	Risk management policy
GP 2.2. Plan the process Address the comprehensive planning for all of the specific practices in this process area, from determining risk sources and categories all the way through to the implementation of risk mitigation plans	Software project management plan or risk management plan
GP 2.3. Provide resources Identification of resources used in support of risk management activities	Software project management plan, or risk management plan
GP 2.4. Assign responsibility Identify individual(s) responsible for risk management	Software project management plan, or risk management plan
GP 2.5. Train people Train individuals responsible for risk management activities	Training plan, software project management plan, or risk management plan
GP 2.6. Manage configurations Description of how work products and all associated artifacts in support of risk management activities are placed under configuration management	Configuration management plan
GP 2.7. Identify relevant stakeholders Describe activities and feedback mechanisms	Software project management plan or risk management plan
GP 2.8. Monitor and control the process Monitor and control, reporting all tailoring and changes associated with risk management	SQA plan and software project management plan
GP 2.9. Objectively evaluate adherence Individuals responsible for SQA activities may conduct reviews	SQA plan and software project management plan
GP 2.10. Review status with higher-level management	These reviews are typically in the form of a briefing presented to the management steering committee by the process group and the process action teams.
These reviews are typically in the form of a briefing presented to the management steering committee by the process group and the process action teams.	Status review briefing

(continued)

Table 6-24. Risk management goals and practices [61]

Specific and generic goals/practices and typical work products	Plan or artifact
GG3. Institutionalize a Defined Process (*cont.*) *GP 3.1. Establish a defined process* Define risk management process activities	Software project management plan or risk management plan
GP 3.2. Collect improvement information	Software measurement and metrics plan, software project management plan, or risk management plan

Risk Management Overview. This section should describe the project's organization structure, its tasks, and its roles and responsibilities. A reference to any associated software project management plan should be provided here.

Risk Management Policies. This section should identify all organizational risk management policies. Refer to IEEE Std 1540-2001 for detailed information supporting risk management requirements.

Organization. This section should depict the organizational structure that assesses and controls any associated project risk. This should include a description of each major element of the organization, together with the roles and delegated responsibilities. The amount of organizational freedom and objectivity to evaluate and monitor the quality of the software, and to verify problem resolutions, should be clearly described and documented. In addition, the organization responsible for preparing and maintaining the risk management plan should be identified.

Table 6-25. Risk management plan document outline

Title Page
Revision Page
Table of Contents
 1. Introduction
 2. Purpose and Scope
 3. Definitions and Acronyms
 4. References
 5. Risk Management Overview
 6. Risk Management Policies
 7. Organization
 8. Responsibilities
 9. Orientation and Training
10. Costs and Schedules
11. Process Description
12. Process Evaluation
13. Communication
14. Plan Management

Responsibilities. This section should identify the specific organizational element that is responsible for performing risk management activities. Individuals that typically participate in risk management activities and should be described are: program manager, project manager, risk management manager, software team members, quality assurance manager, and configuration management manager.

Orientation and Training. This section should describe the orientation or training activities required in support of risk management activities. If this information is supplied in a related training plan, a reference to this plan may be provided here.

Costs and Schedules. Describe all associated cost and schedule impacts in this section. If this information is supplied in a related software project management plan, a reference to this plan may be provided here.

Process Description. Describe the risk management processes employed by the project in this section. If an organizational risk management process exists, then a reference to these processes may be provided here. If any adaptations are required during the adoption of organizational processes at the project level, then these adaptations must be described in this section. A diagram is often useful when communicating process descriptions.

All techniques and tools used during the risk management process should be described. Risk taxonomies, forms, and databases are often used in support of project level risk management activities. If these items are used, a description should be provided.

Process Evaluation. Describe how measurement information will be collected in support of the evaluation of the risk management process. It this information is provided in a related software measurement and metrics plan, and then a reference to this plan may be provided here.

Communication. This section should describe how all associated project risk information would be coordinated and communicated among all project stakeholders. It is important to address when risk elevation procedures are required.

Plan Management. The management of the risk management plan, including revisions, approval, and associated configuration management requirements, should be addressed in this section.

DECISION ANALYSIS AND RESOLUTION

The purpose of Decision Analysis and Resolution (DAR) is to analyze possible decisions using a formal evaluation process that evaluates identified alternatives against established criteria.

DAR belongs in the Support process category for Maturity Level 3, along with Product and Process Quality Assurance, Configuration Management, and Measurement and Analysis, and supports all the process areas by providing a formal evaluation process that ensures that alternatives are compared and the best one is selected to accomplish the goals of the process areas.

DAR addresses formal evaluations that are used in the other process areas, such as when Technical Solutions selects a solution from alternative solutions. DAR involves establishing guidelines to determine which issues should be subjected to a formal evaluation process and then applying formal evaluation processes to these issues. In an immature organization, DAR is frequently a thought process used by experts or managers and is rarely documented for others to understand or to collaborate in. Example decision risk rating matrix and decision tree analyses are provided in support of this PA. Appendix C contains additional information in support of the associated DAR work products in the Cost Benefit Ratio section.

CMMI-SW Goals

SG 1. Decisions are based on an evaluation of alternatives using established criteria.

The organization should make formal decisions on critical issues based on an evaluation of alternatives and using established criteria. Guidelines are used to determine which issues are subject to this formal evaluation process. Criteria are maintained for evaluating alternative solutions, which are selected and evaluated. The solution is selected from the alternatives based on the evaluation criteria.

GG 3. The process is institutionalized as a defined process.

CMMI®-SW specifically requires that the DAR PA be institutionalized as a defined process, so the project team can tailor the processes and document templates from the organization's set of standard processes according to the organization's tailoring guidelines. Table 6-26 describes the specific and generic goals and practices, typical work products, and commonly associated plans or artifacts in support of the PA.

Probability/Impact Risk Rating Matrix

IEEE Std 1490-2003, IEEE Guide—Adoption of the PMI Standard, A guide to the Project Management Body of Knowledge (PMBOK) [34] describes risk management as one of the project management knowledge areas. The PMBOK dedicates an entire chapter to the definition, analysis, and appropriate response to associated project risk. The phrase used by the PMBOK to describe the evaluation of project risk and associated likely outcomes is "risk quantification." Several methods for risk assessment are presented in the PMBOK. These methods range from statistical summaries to expert judgment. The probability/impact risk rating matrix, Figure 6-5, is provided as an example of a way to leverage expert judgment and rating identified project risk.

Decision Tree Analysis

Decision analysis tools and techniques may be used to support the technical decisions that must be made during the lifecycle of a software project. Some of these decisions include the type of architecture, determining whether to build or buy, product design, platform type, life cycle selection, and testing approaches. Table 6-27 provides the steps typical to the decision process and associated sample questions.

Table 6-26. Decision analysis and resolution goals and practices [61]

Specific and generic goals/practices and typical work products	Plan or artifact
SG1 Evaluate Alternatives	
SP 1.1. Establish Guidelines for Decision Analysis	
Guidelines for when to apply a formal evaluation process	Risk management plan, decision or probability/impact risk rating matrix
SP 1.2. Establish Evaluation Criteria	
Documented evaluation criteria	
Rankings of criteria importance	Risk management plan
SP 1.3. Identify Alternative Solutions	
Identified Alternatives	Risk management plan
SP 1.4. Select Evaluation Methods	
Selected Evaluation Methods	Risk management plan, decision tree analysis
SP 1.5. Evaluate Alternatives	
Evaluation Results	Risk management plan, cost/benefit ratio
SP 1.6. Select Solutions	
Recommended solutions to address significant issues	Risk management plan
GG3. Institutionalize a Defined Process	
GP 2.1. Establish an organizational policy	
Establishes organizational expectations for selectively analyzing possible decisions using a formal evaluation process that evaluates identified alternatives against established criteria. The policy should also provide guidance on which decisions require a formal evaluation process.	Organizational policy
GP 2.2. Plan the process	
Typically, this plan for performing the decision analysis and resolution process is included in (or is referenced by) the project plan, which is described in the Project Planning process area.	Software project management plan
GP 2.3. Provide resources	
Identification of resources used in support of decision analysis activities	Software project management plan, or risk management plan
GP 2.4. Assign responsibility	
Identify individual(s) responsible for decision analysis and item identification	Software project management plan, or risk management plan
GP 2.5. Train people	
Train individuals responsible for decision analysis activities	Training plan, project plan, or risk management plan
GP 2.6. Manage configurations	
Description of how work products and all associated artifacts in support of decision analysis activities are placed under configuration management	Configuration management plan

(*continued*)

Table 6-26. (*Continued*)

Specific and generic goals/practices and typical work products	Plan or artifact
GG3. Institutionalize a Defined Process (*cont.*)	
GP 2.7. Identify relevant stakeholders Describe activities and feedback mechanisms	Software project management plan or risk management plan
GP 2.8. Monitor and control the process Monitor and control, reporting all tailoring and changes associated with decision analysis activities	Cost/benefit ratio of using formal evaluation processes
GP 2.9. Objectively evaluate adherence Individuals responsible for SQA activities may conduct reviews	Quality assurance plan and project plan
GP 2.10. Review status with higher-level management These reviews are typically in the form of a briefing presented to the management steering committee by the process group and the process action teams.	Status review briefing
GP 3.1. Establish a defined process Define decision analysis activities	Software project management plan or risk management plan
GP 3.2. Collect improvement information	Software measurement and metrics plan, software project management plan, risk management plan, or decision analysis process

	Consequences				
	Insignificant	Minor	Moderate	Major	Catastrophic
Likelihood	1	2	3	4	5
A (Almost certain)	H	H	E	E	E
B (Likely)	M	H	H	E	E
C (Possible)	L	M	H	E	E
D (Unlikely)	L	L	M	H	E
E (Rare)	L	L	M	H	H

E	Extreme Risk – Immediate action; senior management involved
H	High Risk – Management responsibility should be specified
M	Moderate Risk – Manage by specific monitoring or response
L	Low Risk – Manage by routine process

Figure 6-5. Probability/impact risk rating matrix.

Table 6-27. Basic decision analysis and resolution process [58]

Step in process	Sample questions
1. Draft decision statement	On what situation do we need to take action? What are we trying to achieve?
2. Establish decision objectives	What are the anticipated results? What resources do we have to work with?
3. Objectives: required or desired?	What is critical to the success of the decision and can this be measured?
4. Value the desired objectives	What is the value of each objective? What is the value scale?
5. Develop the alternatives	Are there possible alternative solutions?
6. Test alternatives against required objectives	How do the alternatives compare against the objectives?
7. Score alternatives against desired objectives	What is the value of each alternative? What is the value scale?
8. Determine risks	What are the associated probabilities and impact of the risk?
9. Select best alternative	What alternative provides the most benefit for the least risk?

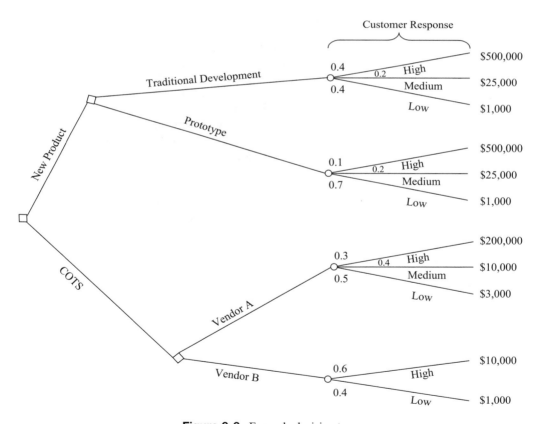

Figure 6-6. Example decision tree.

Table 6-28. Example decision tree results

Node	Benefit level	Benefit calculations	Benefit subtotal	Outcome total	Estimated cost	Net benefit
Traditional	High	0.4 × 500,000	200,000	210,000		
development	Medium	0.4 × 25,000	10,000	75,000		
	Low	0.2 × 1,000	200	135,000		
Prototype	High	0.1 × 500,000	50,000	55,700	40,000	15,700
	Medium	0.2 × 25,000	5,000			
	Low	0.7 × 1,000	700			
Vendor A	High	0.3 × 200,000	60,000	65,500	15,000	50,000
	Medium	0.4 × 10,000	4,000			
	Low	0.5 × 3,000	1,500			
Vendor B	High	0.6 × 10,000	6,000	6,400	0	6,400
	Low	0.4 × 1,000	400			

Decision trees are often useful when attempting to choose between several courses of action. They provide a highly effective structure for the evaluation of options, associated outcomes, and associated risk and benefit. A decision tree begins with a decision that is represented by a small square toward the left of the drawing area. Lines are drawn out to the right from this initial decision box, each line representing an alternative solution.

Consider the results of each decision path. If the alternative presented results in uncertainty, draw a circle, noting the uncertainty above the circle. In a decision tree, squares represent decisions and circles represent uncertain outcomes. If the alternative results in another decision, draw another square, noting the decision above the square. This process should be repeated until all possibilities are exhausted.

The decision tree must now be evaluated. Begin by identifying a percentage value for each of the possible outcomes. Also assign a dollar amount or value to each alternative outcome. Next, perform the calculations associated with each node of the decision tree and record the result. When calculations are based upon uncertainties (circled items) multiply the value of the outcome by the probability shown. After the estimated outcome, or benefit, is calculated, it is important to subtract the estimate for all associated costs. The decision tree will then reflect net benefit values, which may be used in the decision making process. An example of a basic decision tree is presented in Figure 6-6. An example of the recorded outcomes, or benefit, for a new product through development is provided in Table 6-28.

Decision trees can help provide objective insight when analyzing information for decision making. They help to clarify the problem by laying out all the options. All probabilities are factored into the various alternatives. The benefit of an effective decision making tool is that it helps individuals make objective decisions.

7

LEVEL 2 CMMI® FOR SMALL PROJECTS

INTRODUCTION TO CMMI® AND SMALL PROJECTS

Many organizations are intimidated by the amount of documentation associated with the typical CMMI-SW® Level 2. The benefit of CMMI® implementation in support of larger software development efforts is something that has long been recognized and accepted. Project managers embrace the CMMI® for larger software development efforts. Smaller projects* should also strive for process control and improvement and can benefit from it as well. The key to implementing Level 2 controls on a smaller scale is to have the organizational process documentation in place. Also, instead of having all the documentation relate to a single project, the supporting plans (i.e., software requirements management, configuration management, quality assurance, etc.) describe the processes adopted by the managing organization.

In a typical organization, you will find three layers of management: project, program, and organization. To provide support for the small software project, the policies are developed at the organizational level and the supporting plans at the program level. The project is then required to follow these program level plans and need only write its project specific documentation. This approach provides several immediate benefits. It allows the technical manager to get down to the business of managing development, encourages standardization across the organization, and encourages participation in process definition activities at the organizational level.

PROJECT MANAGEMENT PLAN—SMALL PROJECTS

The purpose of the small project plan (SPP) is to help project managers during the development of a small software project. This template is useful for projects with time esti-

---

*Small project is defined here as a project having 3 or fewer individuals, or an operating budget of less than $30,000. Each organization must decide on what it defines as a small project.

mates ranging from one person-month to one person-year. It is designed for projects that strive to follow repeatable and defined processes. This project plan is used to refer to existing organizational policies, plans, and procedures. Exceptions to existing policies, plans, and procedures must be documented with additional information appropriate to each project's SPP.

The SPP provides a convenient way to gather and report the critical information necessary for a small project without incurring the overhead of documentation necessary to support the additional communication channels of a larger project. Software quality assurance (SQA) procedures will be captured in Appendix A of the software development plan (SDP). Appendix B will be used to document software configuration management (SCM) activities. Project management and oversight activities (i.e., risk tracking, problem reporting) will be documented in Appendix C of the SDP. Requirements for the project will be captured in Appendix D of the SPP rather than a separate software requirements specification (SRS). Table 7-1 provides a suggested outline for a SDP.

This small software project management plan template is designed to facilitate the definition of processes and procedures relating to software project management activities.

Table 7-1. Small software project management plan document outline

Title Page
Revision Page
Table of Contents
1. Introduction
    1.1 Identification
    1.2 Scope
    1.3 Document Overview
    1.4 Relationship to Other Plans
2. Acronyms and Definitions
3. References
4. Overview
    4.1 Relationships
    4.2 Source Code
    4.3 Documentation
    4.4 Project Resources
    4.5 Project Constraints
5. Software Process
    5.1 Software Development Process
        5.1.1 Lifecycle Model
    5.2 Software Engineering Activities
        5.2.1 Handling of Critical Requirements
        5.2.2 Recording Rationale
        5.2.3 Computer Hardware Resource Utilization
        5.2.4 Reusable Software
        5.2.5 Software Testing
6. Schedule
Appendix A. Software Quality Assurance
Appendix B. Software Configuration Management
Appendix C. Risk Tracking/ Project Oversight
Appendix D. Software Requirements Specification
Requirements Traceability Matrix

This template was developed using IEEE Std 12207.0 and 12207.1, Standards for Information Technology—Software Life Cycle Processes [39]; IEEE Std 1058-1998, IEEE Standard for Software Project Management Plans[15]; IEEE Std 730-2002, IEEE Standard for Software Quality Assurance Plans [3]; IEEE Std 828-1998, IEEE Standard for Software Configuration Management Plans [4]; IEEE Std 830-1998, IEEE Recommended Practice for Software Requirements Specifications [6]; IEEE Std 1044, Standard Classification for Software Anomalies [13]; IEEE Std 982.1, Standard Dictionary of Measures to Produce Reliable Software [7]; IEEE Std 1045, Standard for Software Productivity Metrics [14]; and IEEE Std 1061, Software Quality Metrics Methodology [16], which have been adapted to support CMMI requirements.

This plan must be supplemented with separate supporting plans. The idea is that this plan may be used to support the development of an SDP for project operating under the same configuration management, quality assurance, risk management, and so on policies and procedures. This eliminates redundant documentation, encourages conformity among development efforts, facilitates process improvement through the use of a common approach, and more readily allows for the cross-project transition of personnel.

Introduction. This section should provide a brief summary description of the project and the purpose of this document.

Identification. This paragraph should briefly identify the system and software to which this SPP applies. It should include such items as identification number(s), title(s), abbreviations(s), version number(s), and release number(s). It should also outline deliverables, special conditions of delivery, or other restrictions/requirements (e.g., delivery media). An example follows:

> This document applies to the software development effort in support of the development of [Software Project] Version [xx]. A detailed development timeline in support of this effort is provided in the supporting project Master Schedule.

Scope. This subsection should briefly state the purpose of the system and software to which this SPP applies. It should describe the general nature of the existing system/software and summarize the history of system development, operation, and maintenance. It should also identify the project sponsor, acquirer, user, developer, support agencies, current and planned operating sites, and relationship of this project to other projects. This overview should not be construed as an official statement of product requirements. It will only provide a brief description of the system/software and will reference detailed product requirements outlined in the SRS, Appendix D. An example follows:

> [Project Name] was developed for the [Customer Information]. [Brief summary explanation of product.]
>
> This overview should not be construed as an official statement of product requirements. It will only provide a brief description of the system/software and will reference detailed product requirements outlined in the [Project Name] SRS.

Document Overview. This subsection should summarize the purpose, contents, and any security or privacy issues that should be considered during the development and implementation of the SPP. It should specify the plans for producing scheduled and unscheduled updates to the SPP and methods for disseminating those updates. It should also explain the mechanisms used to place the initial version of the SPP under change control and to control subsequent changes to the SPP. An example is:

> The purpose of the [Project Name] Software Development Plan (SDP) is to guide [Company Name] project management during the development of [Product Identification]. Software Quality Assurance (SQA) procedures are captured in Appendix A of the SDP. Appendix B is used to document Software Configuration Management (SCM) activities where these activities deviate from [Organization Name] SCM Plan. Project management and oversight activities (i.e., risk tracking, problem reporting) are documented in Appendix C of the SDP when these activities deviate from the [Organization Name] Risk Management Plan. Requirements for the project will be captured in the [Project Name] Software Requirements Specification (SRS). This plan will be placed under configuration management controls according the [Organization Name] Software Configuration Management Plan. Updates to this plan will be handled according to relevant SCM procedures and reviews as described in the [Organization Name] Software Quality Assurance Plan.

Relationship to Other Plans. This subsection should describe the relationship of the SPP to other project management plans. If organizational plans are followed, this should be noted here. An example is:

> There are several other [Organization Name] documents that support the information contained within this plan. These documents include: [Organization Name] Software Configuration Management Plan, [Organization Name] Software Quality Assurance Plan, and [Organization Name] Measurement and Metrics Plan. [Project Name] project specific documentation relating to this plan include: [Project Name] Software Requirements Specifications and [Project Name] Master Schedule. This plan has been developed in accordance with all [Organization Name] software development processes, policies, and procedures.

Acronyms and Definitions. This section should identify acronyms and definitions used within the SPP for each project.

References. This section should identify the specific references used within the SPP. Reference to all associated software project documentation, including the statement of work and any amendments, should be included.

Overview. This section of the SPP should list the items to be delivered to the customer, delivery dates, delivery locations, and quantities required to satisfy the terms of the contract. This list should not be construed as an official statement of product requirements. It will only provide an outline of the system/software requirements and will reference detailed product requirements outlined in the SRS, Appendix D. An example follows:

> The project schedule in support of [Project Name] was initiated [Date] with a target completion date of [Date]. Incremental product deliveries may be requested, but none are identified at this time. The delivery requirements are to install [Project Name] on the current [Project Name] external website. [Project Name] will be delivered once deployment has occurred, and [Customer Name] acceptance of the product. [Customer Name] may request the installation of [Product Name] at one additional location. A [Project Name] user's manual shall also be considered to be required as part of this delivery. Detailed schedule guidance is provided by [document name], located [document location]. Detailed requirements information is provided by [document name], located [document location].

Relationships. This subsection should identify interface requirements and concurrent or critical development efforts that may directly or indirectly affect product development efforts.

Source Code. This subsection should identify source code deliverables. An example is:

> There are no requirements for source code deliverables. If [customer name] requests the source code, it will be delivered on CD-ROM.

Documentation. This subsection should itemize documentation deliverables and their required format. It should also contain, either directly or by reference, the documentation plan for the software project. This documentation plan may either be a separate document, stating how documentation is going to be developed and delivered, or may be included in this plan as references to existing standards with documentation deliverables and schedule detailed therein. An example follows:

> [Project Name] will have on-line help; a user's manual will be created from this online help system. The user's manual will be posted on the [Project Name] website and will be available for download by requesting customers. Please refer to the [Project Name] master schedule for a development timeline of associated users on line help and manual.

Project Resources. This subsection should describe the project's approach by describing the tasks (e.g., req. → design → implementation → test) and efforts (update documentation, etc.) required to successfully complete the project. It should state the nature of each major project function or activity and identify the individuals who are responsible for those functions or activities.

This subsection should also describe the makeup of the team, project roles, and internal management structure of the project. Diagrams may be used to depict the lines of authority, responsibility, and communication within the project. Figure 7-1 is an example of an organizational chart.

The relationships and interfaces between the project team and all nonproject organizations should also be defined by the SPP. Minimum interfaces include those with the customer, elements (management, SQA, SCM, etc.), and other support teams. Relationships with other contracting agencies or organizations outside the scope of the project that will impact the project require special attention within this section.

Figure 7-1. Example project organization.

Project Constraints. This subsection should describe the limits of the project, including interfaces with other projects, the application of the program's SCM and SQA (including any divergence from those plans), and the relationship with the project's customer. This section should also describe the administrative and managerial boundaries between the project and each of the following entities: parent organization, customer organization, subcontracted organizations, or any organizational entities that interact with the project.

This subsection should capture the anticipated volume of the project through quantifiable measurements such as lines of code, function points, number of units modified, number of pages of documentation generated or changed, and so on. It may be useful, if the project is well defined in advance, to break down project activities and perform size estimates on each individual activity. This information can then be tied to the project schedule as defined in Section 5 of the document outline.

Software Development Process. This subsection of the SPP should define the relationships among major project functions and activities by specifying the timing of major milestones, baselines, reviews, work products, project deliverables, and signature approvals. The process model may be described using a combination of graphical and textual notations that include project initiation and project termination activities.

Life Cycle Model. This subsection of the software development process should identify the life-cycle models used for the software development process. It should describe the project organizational structure, organizational boundaries and interfaces, and individual responsibilities for the various software development elements.

Software Engineering Activities. This subsection contains the following five subsections:

> *Handling of Critical Requirements.* This subsection will identify the overall software engineering methodologies to be used during requirements elicitation and system design. An example follows:
>
> > Please refer to the [Organization Name] Software Requirements Management Plan for information regarding the elicitation and management of customer requirements.
>
> *Recording Rationale.* This subsection should define the methodologies used to capture design and implementation decisions. An example is:
>
> > All initial design and implementation decisions will be recorded and posted in the [Project Name] portal project workspace. All formal design documentation will follow the format as described in the [Organization Name] Design Documentation Template.
>
> *Computer Hardware Resource Utilization.* This subsection should describe the approach to be followed for allocating computer hardware resources and monitoring their utilization. An example text is:
>
> > All computer hardware resource utilization is identified in the [Project Name] Spend Plan. Resource utilization is billed hourly as an associated site contract charge. Twenty thousand dollars has been allocated to support the procurement of a development environment. Charges in support of project equipment purchases will be recorded by the Project Manager, [Name], and reported to senior management weekly.
>
> *Reusable Software.* This subsection should describe the approach for identifying, evaluating, and reporting opportunities to develop reusable software products. The following example is an example:

It is the responsibility of each developer to identify reusable code during the development process. Any item identified will be placed in the [Organization Name] Reuse Library that is hosted on the [location]. It is the responsibility of each programmer to determine whether items in the Reuse Library may be employed during [Project Name] development.

Software Testing. This subsection should be further divided to describe the approach for software implementation and unit testing. For example:

This project will use existing templates for software test plans and software unit testing when planning for testing. These separate documents will be hosted [location].

Schedule. This section of the SPP should be used to capture the project's schedule, including milestones and critical paths. Options include Gantt charts (Milestones Etc.™, Microsoft Project™), Pert charts, or simple time lines. Figure 7-2 is an example schedule created with Milestones, Etc.™ for a short (two-month) project.

Appendix A. Software Quality Assurance. This appendix should be divided into the following sections to describe the approach for software quality assurance (SQA): software quality assurance evaluations, software quality assurance records, independence in software quality assurance, and corrective action. This appendix should reference the organizational- or program-level quality assurance policies, plans, and procedures, and should

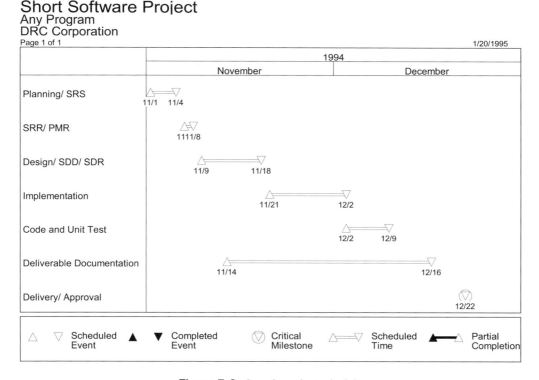

Figure 7-2. Sample project schedule.

detail any deviations from the organizational standard. If a separate SQA plan has been created in support of the development work, simply provide a reference to that document as follows:

> Refer to the [Organization Name] Software Quality Assurance Plan for all applicable processes and procedures.

Appendix B. Software Configuration Management. This appendix should be divided into the following sections to describe the approach for software configuration management (SCM): configuration identification, configuration control; configuration status accounting; configuration audits; and packaging, storage, handling, and delivery. This appendix should reference the organizational- or program-level configuration management policies, plans, and procedures, and should detail any deviations from the organizational standard. If a separate SCM plan has been created in support of the development work, simply provide a reference to that document as follows:

> Refer to the [Organization Name] Software Configuration Management Plan for all applicable processes and procedures.

Appendix C. Risk Tracking/Project Oversight. This appendix should be divided into the following sections to describe the approach for project risk identification and tracking: risk management strategies, measurement activities, and identified project risks. This appendix should reference the organizational- or program-level risk and project management policies, plans, and procedures, and should detail any deviations from the organizational standard. If a separate measurement and metrics plan has been created in support of the development work, simply provide a reference to that document as follows:

> The following measures will be taken as part of project oversight activities. Please refer to the [Organization Name] Software Project Measurement and Metrics Plan for descriptions.
>   (list of measures)

Appendix D. Software Requirements Specification. This appendix is the equivalent of the software requirements specification (SRS) required for large projects. It is produced early in the project life cycle and contains descriptions of all project requirements. This appendix should reference the organizational- or program-level requirements management policies, plans, and procedures, and should detail any deviations from the organizational standard. If a separate SRS has been created in support of the development work, simply provide a reference to that document.

APPENDIX A

IEEE STANDARDS ABSTRACTS

Standard Number	Standard Name	Description
IEEE Std 610.12-1990 (Sep 28) Reaffirmed Sept 2002	IEEE Standard Glossary of Software Engineering Terminology	IEEE Std 610.12-1990, IEEE Standard Glossary of Software Engineering Terminology, identifies terms currently in use in the field of software engineering. Standard definitions for those terms are established.
IEEE Std 730-2002 (Oct 20) Revised Sept 2002	IEEE Standard for Software Quality Assurance Plans	Uniform, minimum acceptable requirements for preparation and content of Software Quality Assurance Plans (SQAPs) are provided. This standard applies to the development and maintenance of critical software. For noncritical software, or for software already developed, a subset of the requirements of this standard may be applied.
IEEE Std 828-1998 (Jun 25)	IEEE Standard for Software Configuration Management Plans	The minimum required contents of a Software Configuration Management Plan (SCMP) are established, and the specific activities to be addressed and their requirements for any portion of a software product's life cycle are defined.
IEEE Std 829-1998 (Sept 16)	IEEE Standard for Software Test Documentation	A set of basic software test documents is described. This standard specifies the form and content of individual test documents. It does not specify the required set of test documents.
IEEE Std 830-1998 (Jun 25)	IEEE Recommended Practice for Software Requirements Specifications	The content and qualities of a good software requirements specification (SRS) are described and several sample SRS outlines are presented. This recommended practice is aimed at specifying requirements of software to be developed but also can be applied to assist in the selection of in-house and commercial software products. *(continued)*

Standard Number	Standard Name	Description
IEEE Std 982.1-1988 (Jun 9)	IEEE Standard Dictionary of Measures to Produce Reliable Software	This standard provides a set of measures indicative of software reliability that can be applied to the software product as well as to the development and support processes. It was motivated by the need of software developers and users who are confronted with a plethora of models, techniques, and measures.
ANSI/IEEE Std 1008-1987 (R1993) Reaffirmed Dec 2002	An American National Standard—IEEE Standard for Software Unit Testing	This standard's primary objective is to specify a standard approach to software unit testing that can be used as a basis for sound software engineering practice.
IEEE Std 1012-1998 (Mar 9)	IEEE Standard for Software Verification and Validation	Software verification and validation (V&V) processes, which determine whether development products of a given activity conform to the requirements of that activity, and whether the software satisfies its intended use and user needs, are described. This determination may include analysis, evaluation, review, inspection, assessment, and testing of software products and processes. V&V processes assess the software in the context of the system, including the operational environment, hardware, interfacing software, operators, and users.
IEEE Std 1012a-1998 (Sept 16)	Supplement to IEEE Standard for Software Verification and Validation: Content Map to IEEE/EIA 12207.1-1996	The relationship between the two sets of requirements on plans for verification and validation of software, found in IEEE Std 1012-1998 and IEEE/EIA 12207.1-1996, is explained so that users may produce documents that comply with both standards .
IEEE Std 1016-1998 (Sept 23)	IEEE Recommended Practice for Software Design Descriptions	The necessary information content and recommendations for an organization for Software Design Descriptions (SDDs) are described. An SDD is a representation of a software system that is used as a medium for communicating software design information. This recommended practice is applicable to paper documents, automated databases, design description languages, or other means of description.
IEEE Std 1028-1997 (Mar 4) Reaffirmed Sept 2002	IEEE Standard for Software Reviews	This standard defines five types of software reviews, together with procedures required for the execution of each review type. This standard is concerned only with the reviews; it does not define procedures for determining the necessity of a review, nor does it specify the disposition of the results of the review. Review types include management reviews, technical reviews, inspections, walk-throughs, and audits.

Standard Number	Standard Name	Description
IEEE Std 1042-1987 Not maintained but available	IEEE Guide to software configuration management	This guide describes the application of configuration management disciplines to management of software engineering projects. This guide serves three groups: developers of software, the software management community, and those responsible for preparation of SCM plans. Software configuration management consists of two major aspects: planning and implementation. This guide focuses on Software configuration management planning and provides broad perspectives for the understanding of software configuration management.
IEEE Std 1044-1993 (Dec 2) Reaffirmed Sept. 2002	IEEE Standard Classification for Software Anomalies	A uniform approach to the classification of anomalies found in software and its documentation is provided. The processing of anomalies discovered during any software life cycle phase are described, and comprehensive lists of software anomaly classifications and related data items that are helpful to identify and track anomalies are provided
IEEE Std 1045-1992 (Sept 17) Reaffirmed Dec 2002	IEEE Standard for Software Productivity Metrics	Consistent ways to measure the elements that go into computing software productivity are defined. Software productivity metrics terminology is given to ensure an understanding of measurement data for both source code and document production.
IEEE Std 1058-1998 (Dec 8)	IEEE Standard for Software Project Management Plans	The format and contents of software project management plans, applicable to any type or size of software project, are described. The elements that should appear in all software project management plans are identified.
IEEE Std 1061-1998 (Dec 8)	IEEE Standard for a Software Quality Metrics Methodology	A methodology for establishing quality requirements and identifying, implementing, analyzing, and validating the process and product software quality metrics is defined. The methodology spans the entire software life cycle.
IEEE Std 1062-1998 Edition (Dec 2) Reaffirmed Sept 2002	IEEE Recommended Practice for Software Acquisition	A set of useful quality practices that can be selected and applied during one or more steps in a software acquisition process is described. This recommended practice can be applied to software that runs on any computer system regardless of the size, complexity, or criticality of the software, but is more suited for use on modified off-the-shelf software and fully developed software.
IEEE Std 1063-2001 (Dec 5)	IEEE Standard for Software User Documentation	Two factors motivated the development of this standard: the concern of the software user communities over the poor quality of much user documentation, and a need for requirements expressed by producers of documentation.

(continued)

Standard Number	Standard Name	Description
IEEE Std 1074-1997 (Dec 9)	IEEE Standard for Developing Software Life Cycle Processes	A process for creating a software life cycle process is provided. Although this standard is directed primarily at the process architect, it is useful to any organization that is responsible for managing and carrying out software projects.
IEEE Std 1175-1991 (Dec 5)	IEEE Standard Reference Model for Computing System Tool Interconnections	The purpose is to establish agreements for information transfer among tools in the contexts of human organization, a computer system platform, and a software development application. Interconnections that must be considered when buying, building, testing, or using computing system tools for specifying behavioral descriptions or requirements of system and software products are described.
IEEE Std 1175.1-2002 (Nov 11)	IEEE Guide for CASoftware Engineering Tool Interconnections— Classification and Description	Introduces and characterizes the problem of interconnecting CASoftware Engineering tools with their environment.
IEEE Std 1219-1998 (Jun 25)	IEEE Standard for Software Maintenance	The process for managing and executing software maintenance activities is described.
IEEE Std 1220-1998 (Dec 8)	IEEE Standard for the Application and Management of the Systems Engineering Process	The interdisciplinary tasks, which are required throughout a system's life cycle to transform customer needs, requirements, and constraints into a system solution, are defined. In addition, the requirements for the systems engineering process and its application throughout the product life cycle are specified. The focus of this standard is on engineering activities necessary to guide product development while ensuring that the product is properly designed to make it affordable to produce, own, operate, maintain, and eventually to dispose of, without undue risk to health or the environment.
IEEE Std 1228-1994 (Mar 17) Reaffirmed Dec 2002	IEEE Standard for Software Safety Plans	The minimum acceptable requirements for the content of a software safety plan are established. This standard applies to the software safety plan used for the development, procurement, maintenance, and retirement of safety-critical software. This standard requires that the plan be prepared within the context of the system safety program. Only the safety aspects of the software are included. This standard does not contain special provisions required for software used in distributed systems or in parallel processors.

Standard Number	Standard Name	Description
IEEE Std 1233, 1998 Edition (Apr 17) Reaffirmed Sept 2002	IEEE Guide for Developing System Requirements Specifications	Guidance for the development of the set of requirements, System Requirements Specification (SyRS) that will satisfy an expressed need is provided. Developing a SyRS includes the identification, organization, presentation, and modification of the requirements. Also addressed are the conditions for incorporating operational concepts, design constraints, and design configuration requirements into the specification. This guide also covers the necessary characteristics and qualities of individual requirements and the set of all requirements.
IEEE Std 1320.1-1998 (Jun 25)	IEEE Standard for Functional Modeling Language—Syntax and Semantics for IDEF0	IDEF0 function modeling is designed to represent the decisions, actions, and activities of an existing or prospective organization or system. IDEF0 may be used to model a wide variety of systems, composed of people, machines, materials, computers, and information of all varieties and structured by the relationships among them, both automated and nonautomated. As the basis of this architecture, IDEF0 may then be used to design an implementation that meets these requirements and performs these functions.
IEEE Std 1320.2-1998 (Jun 25) IEEE Std 1320.2a	IEEE Standard for Conceptual Modeling Language Syntax and Semantics for IDEF1X 97 (IDEF object)	IDEF1X 97 consists of two conceptual modeling languages. The key-style language supports data/information modeling and is downward compatible with the U.S. government's 1993 standard, FIPS PUB 184. The identity-style language is based on the object model with declarative rules and constraints.
IEEE Std 1362-1998 (Mar 19)	IEEE Guide for Information Technology —System Definition— Concept of Operations (ConOps) Document	The format and contents of a concept of operations (ConOps) document are described. A ConOps is a user-oriented document that describes system characteristics for a proposed system from the users' viewpoint. The ConOps document is used to communicate overall quantitative and qualitative system characteristics to the user, buyer, developer, and other organizational elements (for example, training, facilities, staffing, and maintenance). It is used to describe the user organization(s), mission(s), and organizational objectives from an integrated systems point of view.
IEEE Std 1420.1-1995 (Dec 12) Reaffirmed Jun 2002	IEEE Standard for IT Software Reuse—Data Model for Reuse Library Interoperability: Basic Interoperability Data Model (BIDM)	The minimal set of information about assets that reuse libraries should be able to exchange to support interoperability is provided.

(continued)

Standard Number	Standard Name	Description
IEEE Std 1420.1a-1996 (Dec 10) Reaffirmed Jun 2002	Supplement to IEEE Standard for IT Software Reuse—Data Model for Reuse Library Interoperability: Asset Certification Framework	A consistent structure for describing a reuse library's asset certification policy in terms of an asset ertification framework is defined, along with a cstandard interoperability data model for interchange of asset certification information.
IEEE Std 1420.1b-1999 (Jun 26) Reaffirmed Jun 2002	IEEE Trial-Use Supplement to IEEE Standard for IT Software Reuse—Data Model for Reuse Library Interoperability: Intellectual Property Rights Framework	This extension to the Basic Interoperability Data Model (IEEE Std 1420.1-1995) incorporates intellectual property rights issues into software asset descriptions for reuse library interoperability.
IEEE Std 1462-1998 (Mar 19)	IEEE Standard—Adoption of ISO/IEC 14102: 1995—Information Technology—Guideline for the evaluation and selection of CASoftware Engineering tools	ISO/IEC 14102:1995 deals with the evaluation and selection of CASoftware Engineering tools, covering a partial or full portion of the software engineering life cycle. The adoption of the International Standard by IEEE includes an implementation note, which explains terminology differences, identifies related IEEE standards, and provides interpretation of the international standard.
IEEE Std 1465-1998 (Jun 25)	IEEE Standard—Adoption of International Standard ISO/IEC 12119: 1994(E)—Information Technology—Software packages—Quality requirements and testing	Quality requirements for software packages and instructions on how to test a software package against these requirements are established. The requirements apply to software packages as they are offered and delivered, not to the production process (including activities and intermediate products, such as specifications).
IEEE Std 1471-2000 (Sept 21)	IEEE Recommended Practice for Architectural Description of Software Intensive Systems	This recommended practice addressing the activities of the creation, analysis, and sustainment of architectural descriptions. A conceptual framework for architectural description is established. The content of an architectural description is defined. Annexes provide the rationale for key concepts and terminology, the relationships to other standards, and examples of usage.
IEEE Std 1490-1998 (Jun 25)	IEEE Guide—Adoption of PMI Standard—A Guide to the Project Management Body of Knowledge	The subset of the Project Management Body of Knowledge that is generally accepted is identified and described in this guide. "Generally accepted" means that the knowledge and practices described are applicable to most projects most of the time, and that there is widespread consensus about their value and usefulness. It does not mean that the knowledge and practices should be applied uniformly to all projects without considering whether they are appropriate.

Standard Number	Standard Name	Description
IEEE Std 1517-1999 (Jun 26)	IEEE Standard for Information Technology —Software Life Cycle Processes—Reuse Processes	A common framework for extending the software life cycle processes of IEEE/EIA Std 12207.0-1996 to include the systematic practice of software reuse is provided. This standard specifies the processes, activities, and tasks to be applied during each phase of the software life cycle to enable a software product to be constructed from reusable assets. It also specifies the processes, activities, and tasks to enable the identification, construction, maintenance, and management of assets supplied.
IEEE Std 1540-2001 (Mar 17)	IEEE Standard for Software Life Cycle Processes—Risk Management	A process for the management of risk in the life cycle of software is defined. It can be added to the existing set of software life cycle processes defined by the IEEE/EIA 12207 series of standards, or it can be used independently.
IEEE Std 2001-2002 (Jan 21, 2003)	IEEE Recommended Practice for Internet Practices—Web Page Engineering—Intranet/ Extranet Applications	This standard defines recommended practices for Web page engineering. It addresses the needs of webmasters and managers to effectively develop and manage World Wide Web projects (internally via an intranet or in relation to specific communities via an extranet).
IEEE/EIA 12207.0-1996 (Mar)	Standard for Information Technology—Software life cycle processes— Software Life Cycle Processes	ISO/IEC 12207 provides a common framework for developing and managing software. IEEE/EIA 12207.0 consists of the clarifications, additions, and changes accepted by the Institute of Electrical and Electronics Engineers (IEEE) and the Electronic Industries Association (EIA) as formulated by a joint project of the two organizations.
IEEE/EIA 12207.1-1996 (April)	Standard for Information Technology—Software Life Cycle Processes— Software Life Cycle Processes—Life Cycle Data	ISO/IEC 12207 provides a common framework for developing and managing software. IEEE/EIA 12207.0 consists of the clarifications, additions, and changes accepted by the Institute of Electrical and Electronics Engineers (IEEE) and the Electronic Industries Association (EIA) as formulated by a joint project of the two organizations. IEEE/EIA 12207.1 provides guidance for recording life cycle data resulting from the life cycle processes of IEEE/EIA 12207.0.
IEEE/EIA 12207.2-1997 (Apr 1998)	Standard for Information Technology—Software Life Cycle Processes— Software Life Cycle Processes— Implementation Considerations	ISO/IEC 12207 provides a common framework for developing and managing software. IEEE/EIA 12207.0 consists of the clarifications, additions, and changes accepted by the Institute of Electrical and Electronics Engineers (IEEE) and the Electronic Industries Association (EIA) as formulated by a joint project of the two organizations. IEEE/EIA 12207.2 provides implementation consideration guidance for the normative clauses of IEEE/EIA 12207.0. The guidance is based on software industry experience with the life cycle processes presented in IEEE/EIA 12207.0.

(*continued*)

Standard Number	Standard Name	Description
IEEE Std 14143.1-2000 (Jan 30)	IT—Software Measurement— Functional Size Measurement—Part 1: Definition of Concepts	Implementation notes that relate to the IEEE interpretation of ISO/IEC 14143-1:1998 are described.

COMPARISON OF CMMI®-SW LEVELS 2 AND 3 TO IEEE STANDARDS

Cat	CMMI-SW (Levels 2 and 3)	IEEE #	IEEE Standards for:
	ALL PAs	1219-1998	Software Maintenance
	ALL PAs	12207.0-1996	Software Life Cycle Processes
E	Product Integration; Verification; Validation	1012-1998	Software Verification and Validation
E	Requirements Management; Requirements Development	830-1998	Software Requirements Specifications
E	Technical Solution	1016-1998	Software Design Descriptions
E	Technical Solution	1063-2001	Software User Documentation
E	Technical Solution	1471-2000	Architectural Description of Software Intensive Systems
E	Validation	1028-2002	Software Reviews
E	Verification; Validation	829-1998	Software Test Documentation
Pl	Project Planning; Project Monitoring and Control Integrated Product Management;	1490-2003	Project Management Body of Knowledge
Pl	Project Planning; Project Monitoring and Control	1058-1998	Software Project Management Plans
Pl	Risk Management	1540-2001	Risk Management
Pl	Supplier Agreement Management	1062-1998	Software Acquisition
Pr	Organizational Process Definition	1074-1997	Developing Software Life Cycle Processes
Pr	Organizational Process Focus		
Pr	Organizational Training		

(continued)

Practical Support for CMMI®-SW Software Project Documentation. By S. K. Land and J. W. Walz **223**
© 2006 IEEE Computer Society

Cat	CMMI-SW (Levels 2 and 3)	IEEE #	IEEE Standards for:
S	Configuration Management	828-1998	Software Configuration Management Plans
S	Decision Analysis and Resolution	1490-2003	Project Management Body of Knowledge
S	Measurements and Analysis	1045-2002	Software Productivity Metrics
S	Measurements and Analysis	1061-1998	Software Quality Metrics Methodology
S	Process and Product Quality Assurance	730-2002	Software Quality Assurance

Legend: Cat = process categories. E = Engineering, PL = Planning, Pr = Process, and S = Support.

SOFTWARE PROCESS WORK PRODUCTS

REQUIREMENTS MANAGEMENT

Requirements Traceability

Requirements traceability is a CMMI®-SW cornerstone for achieving Level 2, and the traceability throughout the defined life cycle can be accomplished by the addition of a traceability matrix. The following example (Table C-1) supports backward and forward traceability for validation testing. Additional columns should be added to support traceability to software, or system, design and development. The conversion of this type of matrix to a database tracking system is a common practice for developing organizations. There are many commercially available tools that support requirements tracking.

Table C-1. Requirements traceability matrix example

CER #	Requirement name	Priority	Risk	Requirements document paragraph	Validation method(s)	Formal test paragraph	Status
1	Assignments and Terminations Module (A&T) Performance	2	M	3.1.1.1.1.1	Test Inspection Demonstration	4.1.2.3 4.1.2.6 4.1.2.7 4.1.2.9	Open

PROJECT PLANNING

Work-Breakdown Structure

A Work-Breakdown Structure (WBS) defines and breaks down the work associated with a project into manageable parts. It describes all activities that have to occur to accomplish the project. The WBS serves as the foundation for the development of project schedules, budgets, and resource requirements.

A WBS may be structured by project activities or components, functional areas or types of work, or types of resources, and is organized by its smallest component a work package. A work package is defined as a deliverable or product at the lowest level of the WBS. Work packages may also be further subdivided into activities or tasks. IEEE Std 1490-2003, IEEE Guide—Adoption of the PMI Standard, A Guide to the Project Management Body of Knowledge, recommends the use of nouns to represent the "things" in a WBS. Figure C-1 provides an example of a sample WBS organized by activity.

Workflow Diagram

Workflow may be defined as the set of all activities in a project from start to finish. A workflow diagram is a pictorial representation of the operational aspect of a work procedure: how tasks are structured, who performs them, what their relative order is, how they are synchronized, how information flows to support the tasks, and how tasks are being tracked. Workflow problems can be modeled and analyzed using Petri nets. Petri nets are an abstract, formal model of information flow, showing static and dynamic properties of a system [2]. Figure C-2 provides an example of a workflow diagram.

Stakeholder Involvement Matrix [50]

The stakeholder involvement matrix presents a consolidated view of stakeholder involvement. This matrix can be used to define and document stakeholder involvement. Table C-

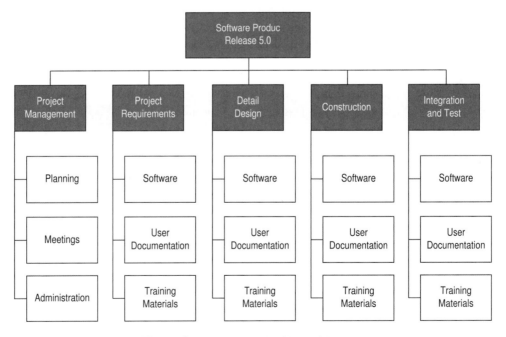

Figure C-1. WBS organized by activity [34].

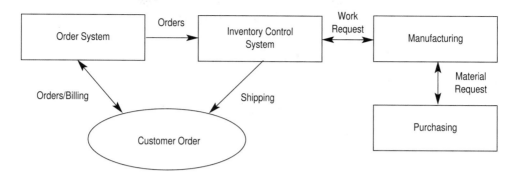

Figure C-2. Example workflow diagram

2 provides an example of a stakeholder involvement matrix. An electronic version of this is provided on the CD-ROM accompanying this book as *Stakeholder Involvement Matrix.doc.*

PROJECT MONITORING AND CONTROL

Open Issues List

The form on the following page (Figure C-3) is presented as an example form that may be used in support of the documentation and tracking of issues associated with a software project. These issues are typically captured during reviews or team meetings and should be reviewed on a regular basis. An electronic version of this work product identified as *Open Issues List.doc* is provided on the CD-ROM accompanying this book.

PROCESS AND PRODUCT QUALITY ASSURANCE

Example Life Cycle

Figure C-4 shows a full development life cycle for a new software product and recommended software quality assurance activities. (Note: This information may be described in this SQA plan or in the associated SPMP.) The following text explains each numbered box in the figure.

Box (0). Tasking is the process of receiving a statement of work (SOW) from the customer. It is the official requirement to which all software development efforts must be directed. The customer, receives requirements from its associated Configuration Control Board (CCB).

Box (1). Software requirements define the problem to be solved by analyzing the customer requirements of the solution in a Software Requirements Specification (SRS). The requirements portion of the life cycle is "the period of time in the software life cycle during which the requirements for a software product are defined and documented" [6].

Customer requirements should be in the form of a Statement of Work, Tasking Statement, or Software System Specification (SSS).

Table C-2. Stakeholder involvement matrix

Project Processes	Customer	Operations	Field Support	Program Management	System Engineering	Software Engineering	System Test	Integration & Test	Integrated Logistics Support	Quality Assurance	Configuration Management	Subcontracts	Help Desk	Suppliers
Proposal Generation	R			L	R	R	R	A	R			R		A
Project Planning	R		R	L	R	R	R	R	R			R		R
Project Monitoring and Control				L	R	R	A	R	R			R		A
Metrics				R	L	R					O			
Risk Management	R			L	R	R	A	A	A					
Requirements Development	R	R	O		L	R	R		A	R				
Operations Telecons	R	R	O	O	L	R	O			O				A
Systems Engineering Review	R	R	R	R	L	R	R	R	R	R	R	R		R
Technical Exchange Meeting	R	A	R	R	L	R	R	A	A	O				A
Interface Control Working Group	R		R	R	L	R	R							A
Software Reviews					O	L	O			O	R			A
SW CCB				O	R	R	R	O		O	R	L		A
HW CCB				O	R		R	R			R	L		
Integration Testing					A	A	L			A				
Validation Testing		L	R		A	A	R			O				
Verification Testing			R				L			O				
Regression Testing			R				L			O				
Site Installation	R		R		A	A	A	L			R	A	A	
Leased Service Support				A	A	A	A	A	A		A	A	L	A
2nd Level Engineering Support					L	R	R		A		A		A	A
Subcontract Mgmt				A	A	A					A		L	A

Legend: R = Required, O = Optional, A = As Needed, L = Leader.

As of: [Date] _____

Identification

This document contains open issues from _____ meeting:

Package	
Author	
Review date	
List version	

Open issues

The matrix below lists any *open issues*, and the response to those issues.
The disposition column indicates the resolution of the issue.

Ref	Severity	Comment	Source	Response	Disposition

Figure C-3. Open issues list.

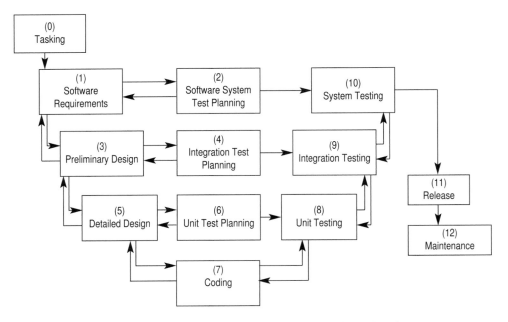

Figure C-4. Full development life cycle. See text for explanation.

The associated Project Lead, Program Manager, and any associated CCB will review the SRS for technical content and clarity. The SRS will be signed by the project Program Manager and a customer representative after joint review. This constitutes their agreement on what is to be produced. The Project Lead, or designee, will review the SRS for adherence to the documentation template provided in Appendix A of the project Software Requirements Management Plan. The SQA Lead will ensure that this review has been completed.

Software Project/Development Planning follows Software Requirements in the Software Development Life Cycle. Software Project/Development Planning requires the documentation of a proposed project's schedule, resources, and manpower estimates. This Software Project Plan will be reviewed for technical content and clarity by the associated Project Lead (PL), Program Manager, senior management, and, if appropriate, the associated CCB. The customer will also participate in Plan review and authorize its implementation.

The PL, or a designee, will also review the Software Project Plan for adherence to the documentation template provided for the company Software Project Plan and checklist. The project Program Manger and a representative from the customer will review actual progress against existing plans to ensure tracking efficiency. These reviews will occur no less than bimonthly. The PM will report progress tracking to company senior management on a monthly basis.

Note. It is important to carefully consider document review coordination time and repair time when project planning.

Box (2). Software System Test Planning follows Software Requirements development in defining how the final product will be tested and what constitutes acceptable results. This step provides the basis for both acceptance and system testing. Issues identified during

Software Test Planning [also referred to as Developmental Test and Evaluation (DT&E) Test Planning] may result in changes to the Software Requirements. A highly desirable by-product of System Test Planning is the further refinement of the Software Requirements. Errors found early in the software development process are less costly and easier to correct.

The Software Test Plan will be reviewed for technical content and clarity by the associated Project Lead, Program Manager, and if appropriate, the associated CCB. The customer will determine, and will agree to, test tolerances. The Project Lead, or a designee, will review the test plan to ensure that all requirements items identified in the SRS are also identified in the Software Test Plan.

Box (3). Preliminary Design breaks the specified problem into manageable components, creating the architecture of the system to be built. The architecture in this step is refined until the lowest level of software components or CSUs are created. The CSUs will represent the algorithms to be used in fulfilling the customer's needs. The preliminary design will be documented as a draft Software Design Document.

The PL will review any existing test plan to ensure that all requirements identified in the SRS are also identified in the Software Design Document. The SQA Lead will ensure that this review has been completed.

Box (4). Integration Test Planning follows Preliminary Design by defining how the components of any related system architecture will be brought back together and successfully integrated. Issues identified during this step will help to define the ordering of component development. Other issues identified during this step include recognition of interfaces that may be difficult to implement. This saves on the final cost by allowing for early detection of a problem. Integration testing will be included as part of the Software (DT&E) Test Plan.

This plan will be reviewed for technical content and clarity by the associated Project Lead, Program Manager, and if appropriate, the CCB. The customer will determine and will agree to test tolerances. The Project Lead, or a designee will review the test plan to ensure that all requirements identified in the SDD are also identified in the Software System Test Plan. The SQA Lead will ensure that this review has been completed.

Box (5). Detailed Design defines the algorithms used within each of the CSUs. The Design is "the period of time in the software life cycle during which the designs for architecture, software components, interfaces, and data are created, documented, and verified to satisfy requirements." [2]

The Detailed Design will be documented as the final version of the Software Design Document (SDD). The Software Design Document will be reviewed for technical content and clarity by the associated Project Lead, Program Manager, and if appropriate, the project CCB. The customer will also participate in SDD review and authorize its implementation.

Box (6). Unit Test Planning follows the Detailed Design by creating the test plan for the individual CSUs. Of primary importance in this step is logical correctness of algorithms and correct handling of problems such as bad data. As in all other test planning steps, feedback to detailed design is essential. Unit testing will be included as part of the Software (DT&E) Test Plan.

The associated Project Lead, Program Manager, and the CCB will review this plan for technical content and clarity. The customer will determine and will agree to test tolerances. The PL will review the test plan to ensure that all requirements items identified in the SRS are also identified in the Software (DT&E) Test Plan. The SQA Lead will ensure that this review has been completed.

Box (7). Coding is the activity of implementing the system in a machine executable form. Implementation is "the period of time in the software life cycle during which a software product is created from documentation and debugged." [2]

The PL, or a designee, using existing coding standards to ensure compliance with design and requirements specifications, will review the Implemented Code. Periodic Program SQA reviews will be conducted to ensure that code reviews are being properly implemented. The results of these reviews will comprise a "lessons learned" document. Results will also be reported to senior management.

Box (8). Unit Testing executes the tests created for each CSU. This level of test will focus on the algorithms with equal emphasis on both "white" and "black" box test cases. White-box testing is testing that is based on an understanding of the internal workings of algorithms. Black-box testing will focus on final results and integrated software operation.

The PL or a designee will "walk through" software test procedures, ensuring that results are recorded and procedures documented based on the evaluation of documented requirements.

Box (9). Integration Testing executes the integration strategy by combining CSUs into higher-level components in an orderly fashion until a full system is available. This level of testing will focus on the interface.

The associated Software/DT&E Test Plan will document system testing procedures and standards.

Box (10). System Testing executes the entire system to ensure the requirements have been satisfied completely and accurately. This level of testing will focus on functionality and is primarily black-box testing. Black-box testing is focused on the functionality of the product (determined by the requirements); therefore, testing is based on input/output knowledge rather than processing and internal decisions.

The associated Software/DT&E Test Plan will document integration testing procedures and standards.

Box (11). Release is the entry of the new or upgraded software into Configuration Management and the replacement of the previous executable with the upgraded executable as appropriate. Please refer to Appendix C for the checklists used in SQA reviews of releasable items.

Box (12). Maintenance is the receipt of and response to problems encountered by the user and upgrades requested by the customer. Each problem or upgrade will spawn a new SQA process. (Refer to Figure C-5.)

Table C3 is a tabular summary of responsibilities for the life cycle process.

The maintenance life cycle of a software project consists of the following stages. The Software Change Process begins with the creation of a change request and is controlled by the Software Configuration Management Process for each program. For this discussion, Change Enhancement Requests (CERs) will include Software Problem Reports (SPRs), Baseline Change Requests (BCRs), Internal Test Reports (ITRs), and Interface Change Notices (ICNs). These forms are shown in the project Software Configuration Management Plan.

The steps in the Software Maintenance Process are shown in Figure C-5. Explanation of the numbered ovals in the figure is provided in the following paragraphs.

Oval (1). Identification and categorization of the CER. This step ensures that the change request is unique (not previously submitted), is valid (not the result of operator error or misunderstanding), and within the scope of the system. Categorization is concerned

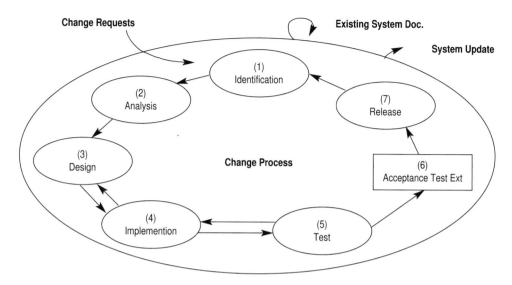

Figure C-5. Software maintenance life cycle. See text for explanation.

Table C-3 Software Development life cycle—responsibility

Life Cycle Activity*	Review by	Product
(0) Tasking	CCB	Work Request ICWG Minutes
(1) Software Requirements	Customer, PM, PL, CCB	SRS SRS Review
(2) Software System Test Planning	Customer, PM, PL, CCB	Software System Test Plan Software System Test Plan Review
Software Project Plan	Customer, PM, PL, CCB	Software Project Plan
(3) Preliminary Design	Customer, PM, PL, CCB	Draft Design Document Draft Design Document Review
(4) Integration Test Plannng	Customer, PM, PL, CCB	Software Test Plan Software Test Plan Review
(5) Detailed Design	Customer, PM, PL, CCB	Software Project Plan SPP Review
(7) Coding	PL, SQAM PL, SQAM	Audit Documentation Walk-through Documentation
(8–10) Unit, Integration, System Testing	Customer, PM, PL, CCB	Software System Test Plan Software System Test Plan Review Documentation
(11) Release	Project Lead	Baseline Release to Configuration Control Inspection Report
(12) Maintenance	Refer Below	Refer Below

*Numbers in parentheses refer to the boxes in Figure C-4).

with the criticality of the change (urgent versus routine) and the type of change (corrective or enhancement).

The Project Lead assumes the initial responsibility for the determination of whether a CER is valid or invalid. Valid change requests require further analysis [oval (2)]. The SQAM will conduct periodic reviews of the Software Configuration Management tracking procedures used for the identification, tracking, and prioritization of CERs.

Oval (2). Analysis examines the source and reason for the change.

The CERs will be reviewed for technical validity and project impact by the associated Project Lead, Program Manager, and, if appropriate, the associated CCB. The customer may also participate in CER review and authorize implementation.

Corrective changes need to be analyzed to find the true source of the problem; enhancements need to be analyzed to determine the scope of the change.

Note: Enhancements are normally additions to the system that entail new or modified requirements. This is similar to starting at the top of Figure xxxx. Part of this step is determining the change to functional testing that will be used upon completion of the change.

Those CERs identified as valid are prioritized and scheduled by the Project Lead, the Program Manager, the Customer, and associated CCBs.

Oval (3). Design retrofit changes into the existing architecture and algorithms. As in development, the integration of changed CSUs and execution of algorithms must be planned early, to catch problems before testing occurs.

The SQAM will conduct periodic reviews to ensure that the Project Lead reviews designs for technical appropriateness. Results of these reviews will be reported to the Program Manager.

Oval (4). Implementation convert new solution into machine executable form. The implemented code will be reviewed using existing coding standards and for compliance to design and requirements specifications. The code will be reviewed by the implementer and then by the PL, or a designee.

Periodic SQA reviews will be conducted to ensure that code reviews are being implemented. Results of these reviews will be reported to the Program Manager.

Oval (5). Testing in this process is the combination of unit, integration, and system testing as described during development. This testing process is often called regression testing due to the need to modify and repeat testing previously performed. A critical issue at this point is the need for repeatability of testing. Changes are vital to the life of the system, as is proper testing; therefore, the testing process should be developed for reuse.

The Project Leads are responsible for ensuring that all CERs scheduled for a production cycle are implemented in the software released for acceptance testing. The SQAM will conduct periodic reviews and the SCM procedures being implemented. Changes to the software baseline will be traced back to the initiating CERs, with any anomalies reported.

Oval (6). Acceptance Testing is carried out to achieve the customer's agreement on completion of the change. Acceptance testing is controlled by the customer for project software products, both for maintenance and development.

Oval (7). Release processes should be considered during identification of the change and should follow an established pattern for the customer. Periodic releases of "builds" are easier to manage and provide more consistency; independent releases provide quicker responses to changing demands. There are considerable trade-offs between the two extremes that should be negotiated with the customer to reach the best mix of speed and control.

The customer and/or associated determines both the "when" and "recipient" of a software release. The PM will, at a minimum, conduct biweekly staff reviews that will in-

Table C-4. Software maintenance life cycle

Life Cycle Activity*	Review by	Product
(1) Identification	PL, PM	CER Categorization SCM Audit
(2) Analysis	PL, PM, CCB, Customer	CER Scheduling Functional Test Review
(3) Design	PL, PM, CCB, Customer Name	Design Review
(4) Implement	PL, PM, SQAM	Walk-through Report Audit Documentation
(5) Test	PL, PM, SQAM	Test Report Audit Documentation
(7) Release	PL, PM	Staff Meeting Minutes

*Numbers in parentheses refer to the ovals in Figure C-5.

clude review of release status and CER implementation progress. The SQA Lead will ensure that these reviews occur as described in the associated software project schedule.

Table C-4 is a tabular summary of the maintenance life cycle process.

Minimum Set of Software Reviews [3]

Table C-5 presents a list of the recommended minimum set of software reviews as described in IEEE Std 730-2002, IEEE Standard for Software Quality Assurance Plans. The list of reviews provided here is not an absolute. Rather, these are provided as an example of what is contained in this standard. Each organization must determine its own recommended minimum set of software reviews.

SQA Inspection Log

An inspector uses the SQA Inspection Log form (Figure C-6) during his/her review of the inspection-package materials. The recorder may also use it during an inspection. In preparing for an inspection, an inspector uses this form to identify all defects found. During an inspection, the recorder uses this form to record all official defects identified. An

Table C-5. Minimum set of software reviews

Software Specification Review
Architectural Design Review
Detailed Design Review
Verification and Validation Plan Review
Functional Audit
Physical Audit
In-Process Audit
Managerial Review
Software Configuration Management Plan Review
Post Implementation Review

```
Log Number:    _____
      Date:_____
Software
Item:_____
Inspector:     _____
Moderator:     _____
Author:        _____
Time spent on review : _____
Review Type:   _____
```

Severity	Defect Description/Class	Category

Figure C-6. SQA Inspection Log Form.

electronic version of this log form, entitled *SQA Inspection Log.doc,* is provided on the CD-ROM accompanying this book.

The following list describes the terms in the form:

Log Number. The log or tracking number assigned to this inspection.

Date. During an inspection, the recorder enters the date on which the inspection occurred. The Date block can be ignored by an inspector using this form during the preinspection review.

Software Item. The name of the software item being inspected.

Inspector. The Inspector block can be left blank when used by the recorder during an inspection. When using this form during a preinspection review, the inspector name is entered in this block.

Moderator. The moderator's name.

Author. The name of the author responsible for the inspection package is entered in this block

Review Type. The type of inspection being conducted is entered in this block.

Category. A category is a two-character code that specifies the class and severity of a defect. A category has the following format: XY, where X is the severity (Major/mInor) and Y is the class.

Severity. Major = a defect that would cause a problem in program operation; mInor = all other defects.

Class. M(issing) = required item is missing from software element; S(tandards Compliance) = nonconformance to existing standards; W(rong) = an error in a software element; E(xtra) = unneeded item is included in a software element; A(mbiguous) = an item of a software element is ambiguous; I(nconsistent) = an item of a software element is inconsistent.

CONFIGURATION MANAGEMENT

Configuration Control Board (CCB) Letter of Authorization

An example of a formal letter of CCB authorization is provided in Figure C-7. This letter may be used to help officially sanction CCB groups. An electronic version of this letter, entitled *CCB Letter of Authorization.doc,* is provided on the CD-ROM accompanying this book.

Configuration Control Board Charter

The following is provided as the suggested example content and format of a Configuration Control Board (CCB) charter. This charter may be used to establish project oversight, control, and responsibility. An electronic version of this charter example, entitled *CCB Charter.doc,* is provided on the CD-ROM accompanying this book.

SCOPE. This Configuration Control Board (CCB) Charter establishes guidelines for those participating in the [Project Name] CCB process. Any questions regarding this document may be directed to [Chartering Organization/Customer].

CCB Overview. The [Project Name] CCB is the management organization responsible for ensuring the documentation and control of [Project Name] software development by ensuring proper establishment, documentation and tracking of system requirements. In addition, the CCB serves as the forum to coordinate software changes between represented agencies and assess the impacts caused by these changes.

CCB Meetings. The [Project Name] CCB will be conducted quarterly and will include the following agenda items:

a. Recommended changes to the CCB Charter

b. (Project Name) schedule

c. Discussion of proposed Baseline Change Requests (BCRs) for approval/disapproval

---

MEMORANDUM FOR [Approving Authority]

DATE: [Date]

FROM: [Company Name]
          [Program Name]
          [Name, Program Manager]
          [Name, Project Manager

SUBJECT: Establishment of [Project Name] Configuration Control Board (CCB)

1. This is to formally propose the charter of the [Project Name] Configuration Control Board (CCB).

2. It is our desire that the [Project Name] CCB be chartered (Attach.1) with the authority to act as the management organization responsible for ensuring the documentation and control of [Project Name] software development by ensuring proper establishment, documentation and tracking of system requirements. In addition, the CCB will serve as the forum to coordinate software changes between represented agencies and assess the impacts caused by requirements changes.

3. Those recommended to participate in CCB voting will be [Approving Authority] housing representatives from the following:

        Management Board:
        [Organization Name]

        Beta Test Site [Organization Name] Representatives:
        [List of sites]

        Contractor Associates:
        [Company Name]
        [Subcontractor Name]

4. The CCB will be the necessary "single voice," providing clear direction during [Project Name] requirements definition, product design and system development. It is our goal to produce the highest quality product for the [Project Name] user. It is our desire that the [Project Name] CCB be chartered to provide [Project Name] development with consolidated and coordinated program direction.

Figure C-7. Example letter of authorization.

    d. Review of new BCRs

    e. Status of open action items

    f. Review of new action items

    g. Program risk assessment

Meeting dates and times will be determined by CCB consensus and announced at least 60 days prior to any scheduled meeting. Reminders will be distributed via DoD Messages, electronic mail, memoranda, and/or bulletin board announcements.

Formal minutes will be available and distributed within 30 days of the meeting's conclusion.

A CCB report will be prepared and maintained by the CCB chairman, listing CCB membership; individual representatives; interfacing modules, subsystems, or segments; pending or approved BCRs; status of software development activities such as open action items and risk tracking; and any other identified pertinent [Project Name] program information.

Figure C-8 provides an example of a BCR. The remainder of this section provides an explanation of the CCB.

Participation. CCB participation is open to all government and contractor organizations that develop or use the project's associated systems or elements. Participants belong to one of six categories:

1. Management Board (MB)
2. Chairman
3. Board Members (Command Representatives)
4. Implementing Members
5. Associate Members
6. Advisors

Originator	Module:		
Name:	Title:		BCR No.:
Organization:	SRS Revision:		Date:
BCR Title:			
Change Summary:			
Reason For Change:			
Impact if Change Not Made:			

Org	Name	Yes	No	Desired Applicability
				Release Available:
				Release Unsupported:
				CCB Approval
				Name:
				Date:
				Signature:
				_____ Approved
				_____Disapproved
				Comments:

Figure C-8. Example Baseline Change Request (BCR).

Discussions concerning the project's BCRs will be limited to the organizations responsible for the implementation of the BCRs. If other organizations are interested in the operational usage of the BCRs, these issues should be brought to the attention of the Board Members.

Participants. The MB is composed of three or less people at a very high level with the ability to make command decisions for the CCB. The purpose of the MB is to address software development problems that cannot be resolved at the CCB Board level. The MB will be chaired by _____ and co-chaired by _____. The chairman: _____ is responsible for establishing and running the CCBs in accordance with this plan, including the following:

a. Leading the effort to document and control all module items associated with HIMS.
b. Ensuring that a proper forum exists for coordinating BCRs and selecting the solution that is in the best interest of the client.
c. Ensuring that minutes and action items are recorded and distributed.

The Chairman has the final approval authority for all BCRs.
Board Members are responsible for the following:

a. Reviewing BCRs to ensure they support operational requirements.
b. Voting to approve/disapprove BCRs.

Implementing Members are responsible for the following:

a. Implementing BCRs.
b. Reviewing BCRs for adequacy and program impact (cost or schedule).

Associate Members are responsible for BCRs for adequacy.
Advisors are responsible for presenting information relevant to a BCR when requested by the sponsoring Operational, Implementing, or Associate Member.

CCB Process

General. All CCB actions are coordinated and take place through the BCR process. Any CCB Participant may propose BCRs, but proprietary BCRs will not be accepted or discussed. It is important to understand that the project's technical interchange meetings (TIMs) provide the forum to brainstorm and discuss preliminary BCR ideas prior to the formal process described in this document. The CCB acts only as the final review and adjudication for BCRs.

Software Requirements Specifications (SRSs). The project's software requirements are brought under CCB control for the BCR process. A BCR describing the request for a software requirements baseline is generated and distributed at the CCB. The BCR is reviewed using the accepted practices and, when approved, the SRS is accepted for CCB management.

BCR Format. BCRs are prepared by generating "is" pages and attaching those to a BCR cover sheet to the document. "Is" pages are change pages indicating how the page should read.

BCR Process. BCRs can be coordinated using any of the following methods:

a. BCRs will be provided to _____ at least thirty (30) days prior to a CCB meeting so they may be reproduced and included in the agenda. Please fax to: _____. *The BCRs should be sent to the Chair and distributed to all participants.)

b. The Board, Implementing, or Associate Member sponsoring the BCR will be provided the opportunity to present a five-minute overview of why the change is needed and what is being changed.

c. Any Board, Implementing, or Associate Member potentially impacted by the BCR will identify himself/herself as a reviewing member for the BCR.

d. Reviewing members will submit their concurrence or nonconcurrence to the sponsoring member and Chair. (Silence is considered concurrence.) If a reviewing member nonconcurs, any issues associated with implementing the BCR, including alternate solutions, will be documented and sent to the sponsoring member.

e. If issues are identified, the sponsoring member will work with all reviewing members to try and resolve conflicts prior to discussion at the CCB.

f. At the CCB, each potential BCR is reviewed for concurrence/nonconcurrence. BCRs that all reviewing agencies concur upon will be signed by the Board members and approved by the Chairman. These BCRs will be prioritized and scheduled for implementation.

g. BCRs that have received nonconcurs go through a formal review. The formal review consists of each reviewing member presenting a 10-minute presentation explaining their concurrence or nonconcurrence with the BCR, including their preferred approach and contractual, cost, and schedule impacts.

h. The Chairman will decide whether to approve, disapprove, or defer the BCR. If the BCR is deferred, the process is repeated again starting with "d" above and the reviewing agencies will use this time to attempt to reach concurrence.

Accelerated Process. (Emergency BCR):

a. BCRs will be provided to _____ and all affected CCB members at least 45 days prior to the CCB. Reviewing members will submit their concurrence or nonconcurrence to the sponsoring member and Chair. (Silence is considered concurrence.)

b. At the CCB, each potential BCR is reviewed for concurrence/nonconcurrence. BCRs that all reviewing agencies concur upon will be signed by Implementing Members and approved by the Chairman.

c. BCRs that have received nonconcurs will go through a formal review. The formal review consists of each reviewing member presenting a 10-minute presentation explaining why they concur or nonconcur with the BCR, including their preferred approach and contractual, cost, and schedule impacts.

d. The Chairman will decide whether to approve, disapprove, or defer the BCR. If the BCR is deferred, the process is repeated starting with "a" above. The reviewing agencies will use this time to attempt to reach concurrence.

Contractual Direction. After a decision has been made by the CCB, all Implementing Members affected by the change will take the contractual action needed to ensure that the change is implemented.

Software Change Request Procedures

A change to software may be requested by one of the following change enhancement requests (CER):

Internal Test Report (ITR)
Software Problem Report (SPR)
Baseline Change Request (BCR)

from external or internal sources or from the product specification/requirements. A Document Change Request (DCR) may be used to request a change to documentation. All of the above requested change procedures are referred to as a change request in the text below. These procedures are outlined below for the project's maintenance and production lifecycles. The procedure is shown graphically in Figure C-9.

The Configuration Control Board (CCB) for each development team determines which of the change requests is required for each software release prior to the start of work for that release. Additional change requests are reviewed by the project's CCB to determine and assign the proper status to the change requests. These are held for CCB scheduling. Status is one of the following:

Open   Change request is to be implemented for current software version.

Hold   Change request targeted for another software version.

Voided   Change request is a duplicate of an existing CER or does not apply to existing software.

Working   Change is currently being implemented.

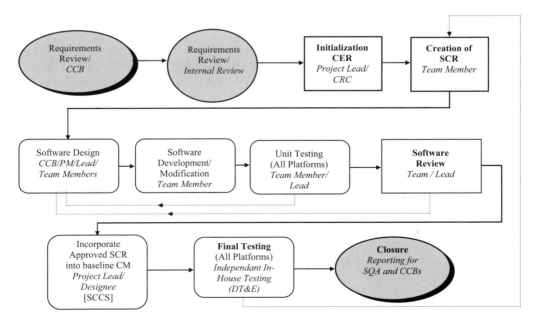

Figure C-9. Software baseline change process.

Testing   Implemented change request is under DT&E evaluation.

Fixed   Change request is implemented, unit and integration test complete.

The following steps define the procedure for each status. Each procedure starts with the change request being given to the Project Lead or designated Change Request Coordinator (CRC).

A. Open
1. INITIALIZATION
   a) The CRC for the associated project creates a CER.
   b) The CRC notifies the Project Lead of the new CER.
2. CREATE SCR
   a) The PL reviews the CER, and provides estimates of schedule impact to CCB.
   b) CCB prioritizes CER and authorizes implementation.
   c) The PL assigns the CER to a Software Engineer (SE) to incorporate.
   d) The SE creates a Software Change Request (SCR).
3. SOFTWARE DESIGN
   The SE designs changes necessary to implement CER:
   a) The SE identifies interface changes and consults with appropriate PLs.
   b) Determines changes needed.
   c) The SE identifies any changes needed for documentation.
   d) The SE estimates work effort required for CER completion.
4. SOFTWARE DEVELOPMENT/MODIFICATION
   a) The SE makes needed changes and provides updated documentation for testing.
   b) The SE checks out the files from the configuration managed version (not the baselined version) of all the files needed to incorporate the CER.
   c) The SE fills in the CM sections of the SCR.
   d) The SE puts the changes in the configuration managed files.
5. UNIT TESTING
   The SE tests on all platforms. If test fails go back to Step 3, Software Design.
6. SOFTWARE REVIEW
   a) The SE prepares a software review package:
      1. Hard copy of SCR.
      2. List of documentation changes.
   b) Software Review
      1. SE gives software review package to another team member for review.
      2. If SCR peer approved go to Step 7.
      3. SCR not approved go back to Step 3, Software Design.
7. INCORPORATE APPROVED SCR INTO BASELINE CM
   a) The PL changes the SCR status to indicate that it is approved.
   b) The SE checks in the files.
   c) The SE fills in the CM sections of the SCR.
   d) The CRC reviews SCR for completeness.
   e) The CRC changes the status on the CER to indicate testing.
   f) The designated SCM gives a list of changed files to PLs for updating their working directories/executables.
   g) The designated SCM installs changes to baseline on all platforms.
8. FINAL TESTING
   a) The CRC identifies all CERs with testing status.

    b) Final testing by DT&E validation group. (If problems are found, start over at Step 2a)

9. CLOSURE

The CRC changes final status on CER to indicate complete.

B. Hold

The CRC creates a CER marking status HOLD and indicates the targeted version for incorporating the change. The CRC files the original change request for the next version. The change will be considered for inclusion in the next version.

C. Voided

1. Duplicates:

    a) The CRC marks the change requests as a duplicate, specifying which CER is a duplicated.

    b) The CRC files the change request for associated version.

2. Other:

    a) The CRC marks the reason for cancellation.

    b) The CRC files the change request for associated version.

D. Testing

The CRC forwards CER/SCR to PL for review. After review, the PL forwards CER to DT&E test for validation. CER status is updated to indicate testing status.

E. Fixed

The CRC files the closed change request and updates database. Figure C-10 provides an overview of the software change request procedures.

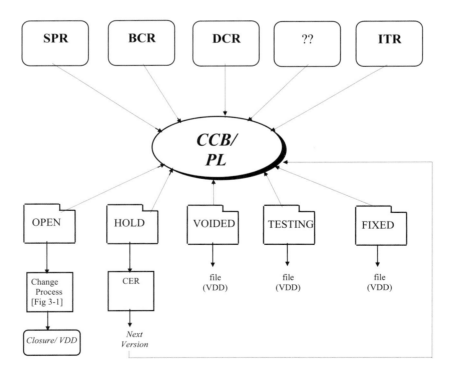

Figure C-10. Software change request procedures.

SUPPLIER AGREEMENT MANAGEMENT

Recommendations for Software Acquisition

IEEE Std 1062-1998, IEEE Recommended Practice for Software Acquisition, describes the software acquisition process. IEEE Std 12207.0 also recommends a set of objectives in support of the Acquisition process. Table C-8 describes these combined recommendations and should be used to support the complete life cycle of the acquisition process.

Table C-9. lists the acquisition checklists provided in IEEE Std 1062-1998. Items listed also support conformance with IEEE/EIA 12207.0-1996. Items in boldface are included in this appendix.

Table C-8. Nine recommendations for the software acquisition process

1	Planning organizational strategy. Review acquirer's objectives and develop a strategy for acquiring software [17].
2	Implementing organization's process. Establish a software acquisition process that fits the organization's needs for obtaining a quality software product. Include appropriate contracting practices [17].
3	Determining the software requirements. Define the software being acquired and prepare quality and maintenance plans for accepting software supplied by the supplier [17].
4	Identifying potential suppliers. Select potential candidates who will provide documentation for their software, demonstrate their software, and provide formal proposals. Failure to perform any of these actions constitutes ground for rejection of a potential supplier. Review supplier performance data from previous contracts [17].
5	Preparing contract requirements. Describe the quality of the work to be done in terms of acceptable performance and acceptance criteria, and prepare contract provisions that tie payments to deliverables [17]. Develop a contract that clearly expresses the expectation, responsibilities, and liabilities of both the acquirer and the supplier. Review contract with legal counsel [39]
6	Evaluating proposals and selecting the supplier. Evaluate supplier proposals, select a qualified supplier, and negotiate the contract. Negotiate with an alternate supplier if necessary [17]. Obtain products and/or services that satisfy the customer need. Qualify potential suppliers through an assessment of their capability to produce the required software [39].
7	Managing supplier performance. Monitor supplier's progress to ensure all milestones are met and to approve work segments. Provide all acquirer deliverables to the supplier when required [17]. Manage the acquisition so that specified constraints and goals are met. Establish a statement of work to be performed under contract. Regularly exchange progress information with the supplier [39].
8	Accepting the software. Perform adequate testing and establish a process for certifying that all discrepancies have been corrected and that all acceptance criteria have been satisfied [17], [39].
9	Using the software. Conduct a follow-up analysis of the software acquisition contract to evaluate contracting practices, record lessons learned, and evaluate user satisfaction with the product. Retain supplier performance data [17]. Establish a means by which the acquirer will assume responsibility for the acquired software product or service [39].

Table C-9. IEEE Std 1062-1998
acquisition checklist summary

Checklist 1:	**Organizational strategy**
Checklist 2:	Define the software
Checklist 3:	**Supplier evaluation**
Checklist 4:	Supplier and acquirer obligations
Checklist 5:	Quality and maintenance plans
Checklist 6:	User survey
Checklist 7:	**Supplier performance standards**
Checklist 8:	Contract payments
Checklist 9:	Monitor supplier progress
Checklist 10:	Software evaluation
Checklist 11:	Software test
Checklist 12:	Software acceptance

Organizational Acquisition Strategy Checklist

IEEE Std 1062-1998, IEEE Recommended Practice for Software Acquisition, provides a example checklist in support of the definition of an organizational acquisition strategy as described in Table C-10. Any checklist used by an organization to determine acquisition strategy should reflect the requirements of the acquiring organization. An electronic version of this document, entitled *Acquisition Strategy Checklist.doc,* is provided on the CD-ROM accompanying this book.

Supplier Evaluation Criteria

IEEE Std 1062-1998, IEEE Recommended Practice for Software Acquisition, provides a checklist in support supplier evaluation. This evaluation criterion is provided in Table C-11. The information provided here provides an illustrative example and should be tailored to reflect organizational process needs. An electronic version of this checklist, entitled *Supplier Checklist.doc,* is provided on the CD-ROM accompanying this book.

Supplier Performance Standards

IEEE Std 1062-1998, IEEE Recommended Practice for Software Acquisition, provides a checklist in support of determining satisfactory supplier performance. This evaluation, provided in Table C-12, is provided as an example and should be based upon all known requirements and constraints unique to the development effort. An electronic version of this checklist, identified as *Supplier Performance Standards.doc,* is provided on the CD-ROM accompanying this book.

MEASUREMENT AND ANALYSIS

List of Measures for Reliable Software [7]

Table C-13 provides a list of suggested measures in support of the production of reliable software. This list is a reproduction of content found in IEEE Std 982.1, IEEE Standard Dictionary of Measures to Produce Reliable Software.

Table C-10. Acquisition strategy checklist [17]

1. Who will provide software support?	Supplier ☐	Acquirer ☐
2. Is maintenance documentation necessary?	Yes ☐	No ☐
3. Will user training be provided by the supplier?	Yes ☐	No ☐
4. Will acquirer's personnel need training?	Yes ☐	No ☐
5. When software conversion or modification is planned:		
a. Will supplier manuals sufficiently describe the supplier's software?	Yes ☐	No ☐
b. Will specification be necessary to describe the conversion or modification requirements and the implementation details of the conversion or modification?	Yes ☐	No ☐
c. Who will provide these specifications?	Supplier ☐	Acquirer ☐
d. Who should approve these specifications? _____		
6. Will source code be provided by the supplier so that modifications can be made?	Yes ☐	No ☐
7. Are supplier publications suitable for end users?	Yes ☐	No ☐
a. Will unique publications be necessary?	Yes ☐	No ☐
b. Will unique publications require formal acceptance?	Yes ☐	No ☐
c. Are there copyright or royalty issues?	Yes ☐	No ☐
8. Will the software be evaluated and certified?	Yes ☐	No ☐
a. Is a survey of the supplier's existing customers sufficient?	Yes ☐	No ☐
b. Are reviews and audits desirable?	Yes ☐	No ☐
c. Is a testing period preferable to demonstrate that the software and its associated documentation are usable in their intended environment?	Yes ☐	No ☐
d. Where will the testing be performed? _____		
e. Who will perform the testing? _____		
f. When will the software be ready for acceptance? _____		
9. Will supplier support be necessary during initial installations of the software by the end users?	Yes ☐	No ☐
10. Will subsequent releases of the software be made?	Yes ☐	No ☐
a. If so, how many? _____ compatible upgrades?	Yes ☐	No ☐
11. Will the acquired software require rework whenever operating system changes occur?	Yes ☐	No ☐
a. If so, how will the rework be accomplished? _____		
12. Will the acquired software commit the acquiring organization to a software product that could possibly be discontinued?	Yes ☐	No ☐
13. What are the risks/options if software is not required? _____		

Example Measures

Management Category

Manpower Metric. Manpower metrics are used primarily for project management and do not necessarily have a direct relationship to other technical and maturity metrics. This metric should be used in conjunction with the development progress metric. The value of this metric is somewhat tied to the accuracy of the project plan, as well as to the accuracy of the labor reporting.

This measure provides an indication of the application of human resources to the development program and the ability to maintain sufficient staffing to complete the project. It can also provide indications of possible problems with meeting schedules and staying

Table C-11. Supplier evaluation checklist [17]

Financial soundness
1. Can a current financial statement be obtained for examination?
2. Is an independent financial rating available?
3. Has the company or any of its principals ever been involved in bankruptcy or litigation?
4. How long has the company been in business?
5. What is the company's history?

Experience and capabilities
1. On a separate page, list by job function the number of people in the company.
2. On a separate page, list the names of sales and technical representatives and support personnel. Can they be interviewed?
3. List the supplier's software products that are sold and the number of installations of each.
4. Is a list of users available?

Development and control processes
1. Are software development practices and standards used?
2. Are software development practices and standards adequate?
3. Are the currently used practices written down?
4. Are documentation guidelines available?
5. How is testing accomplished?

Technical assistance
1. What assistance is provided at the time of installation?
2. Can staff training be conducted on-site?
3. To what extent can the software and documentation be modified to meet user requirements?
4. Who will make changes to the software and documentation?
5. Will modification invalidate the warranty?
6. Are any enhancements planned or in process?
7. Will future enhancement be made available?

Quality practices
1. Are the development and control processes followed?
2. Are requirements, design, and code reviews used?
3. If requirements, design, and code reviews are used, are they effective?
4. Is a total quality program in place?
5. If a total quality program is in place, is it documented?
6. Does the quality program ensure that the product meets specifications?
7. Is a corrective action process established to handle error corrections and technical questions?
8. Is a configuration management process established?

Maintenance service
1. Is there a guarantee in writing about the level and quality of maintenance services provided?
2. Will ongoing updates and error conditions with appropriate documentation be supplied?
3. Who will implement the updates and error corrections?
4. How and where will the updates be implemented?
5. What turnaround time can be expected for corrections?

Product usage
1. Can a demonstration of the software be made at the user site?
2. Are there restrictions on the purposes for which the product may be used?
3. What is the delay between order placement and product delivery?
4. Can documentation be obtained for examination?
5. How many versions of the software are there?
Are error corrections and enhancements release dependent?

Table C-11. *Continued*

Product warranty

1. Is there an unconditional warranty period?
2. If not, is there a warranty?
3. Does successful execution of an agreed-upon acceptance test initiate the warranty period?
4. Does the warranty period provide for a specified level of software product performance for a given period at the premises where it is installed?
5. How long is the warranty period?

Costs

1. What pricing arrangements are available?
2. What are the license terms and renewal provisions?
3. What is included in the acquisition price or license fee?
4. What costs, if any, are associated with an unconditional warranty period?
5. What is the cost of maintenance after the warranty period?
6. What are the costs of modifications?
7. What is the cost of enhancement?
8. Are updates and error corrections provided at no cost?

Contracts

1. Is a standard contract used?
2. Can a contract be obtained for examination?
3. Are contract terms negotiable?
4. Are there royalty issues?

What objections, if any, are there to attaching a copy of these questions with responses to a contract?

---

within budget. It is used to examine the various elements involved in staffing a software project. These elements include the planned level of effort, the actual level of effort, and the losses in the software staff measured per labor category. Planned manpower profiles can be derived from the associated project planning documents.

The planned level of effort is the number of labor hours estimated to be worked on a software module during each tasking cycle. The planned levels are monitored to ensure that the project is meeting the necessary staffing criteria.

Life Cycle Application. The shape of the staff profile trend curve tends to start at a moderate level at the beginning of the contract, grow through design, peak at coding/testing, and diminish near the completion of integration testing. Individual labor categories, however, are likely to peak at different points in the life cycle. The mature result would show little deviation between the planned and actual levels for the entire length of development and scheduled maintenance life cycles.

Algorithm/Graphical Display. The metric graphical display should reflect, by project deliverable software item, the actual versus planned labor hours of effort per month. This metric can be further refined by breaking labor hours into categories (i.e., experienced, novice, special).

Significant deviations (those greater than 10%) of actual from planned levels indicate potential problems with staffing. Deviations between actual and planned levels can be de-

Table C-12. Supplier performance standards [17]

Performance criteria
1. Approach to meet software's functional requirements is defined
2. Growth potential or expansion requirement of the system is defined.
3. Supplier meets time constraints for deliverables.
4. Test and acceptance criteria that are to be met are defined.
5. Programming language standards and practices to be followed are defined.
6. Documentation standards to be followed are defined.
7. Ease of modification is addressed.
8. Maximum computer resources allowed, such as memory size and number of terminals, are defined.
9. Throughput requirements are defined.

Evaluation and test
1. Software possesses all the functional capabilities required.
2. Software performs each functional capability as verified by the following method(s): documentation evaluation, demonstration, user survey, test.
3. Software errors revealed are documented.
4. Software performs all system-level capabilities as verified by a system test.

Correction of discrepancies
1. Supplier documents all identified discrepancies.
2. Supplier establishes discrepancy correction and reporting.
3. Supplier indicates warranty provisions for providing prompt and appropriate corrections.

Acceptance criteria
1. All discrepancies are corrected.
2. Prompt and appropriate corrections are provided.
3. Satisfactory compliance to contract specifications is demonstrated by evaluations and test.
4. Satisfactory compliance to contract specifications is demonstrated by field tests.
5. All deliverable items are provided.
6. Corrective procedures are established for correction of errors found after delivery.
7. Satisfactory training is provided.
8. Satisfactory assistance during initial installation(s) is provided.

---

tected, analyzed, and corrected before they negatively impact the development schedule. Losses in staff can be monitored to detect a growing trend or significant loss of experienced staff. This indicator assists in determining whether there have been a sufficient number of employees to produce the product in the tasking cycle.

Data Requirements. Software tasking documentation must reflect expected staffing profiles and labor hour allocations. For each labor category tracked, document the following items:

1. Labor category name:
   Experienced
      Programmer
      Lead
   Senior
      Support

Table C-13. List of suggested measures for reliable software

Paragraph	Description of Measure
4.1	Fault Density
4.2	Defect Density
4.3	Cumulative Failure Profile
4.4	Fault-Days, Number
4.5	Functional or Modular Test Coverage
4.6	Cause and Effect Graphics
4.7	Requirements Traceability
4.8	Defect Indices
4.9	Error Distribution
4.10	Software Maturity Index
4.11	Man Hours per Major Defect Detected
4.12	Number of Conflicting Requirements
4.13	Number of Entries and Exits per Module
4.14	Software Science Measures
4.15	Graph-Theoretic Complexity for Architecture
4.16	Cyclomatic Complexity
4.17	Minimal Unit Test Case Determination
4.18	Run Reliability
4.19	Design Structure
4.20	Mean Time to Discover the Next K Faults
4.21	Software Purity Level
4.22	Estimated Number of Faults Remaining
4.23	Requirement Compliance
4.24	Test Coverage
4.25	Data or Information Flow Complexity
4.26	Reliability Growth Function
4.27	Residual Fault Count
4.28	Failure Analysis Using Elapsed Time
4.29	Testing Sufficiency
4.30	Mean Time to Failure
4.31	Failure Rate
4.32	Software Documentation and Source Listings
4.33	Required Software Reliability
4.34	Software Release Readiness
4.35	Completeness
4.36	Test Accuracy
4.37	System Performance Reliability
4.38	Independent Process Reliability
4.39	Combined Hardware and Software Operational Availability

2. For each experience level per tasking cycle:
   Number of planned personnel staffing
   Number of personnel actually on staff in current reporting period
   Number of unplanned labor hour losses
   Number of labor hours that are planned to be expended in next reporting period
     (cumulative)
   Number of labor hours that are actually expended in the current reporting period
     (cumulative)

Frequency and Type of Reporting. There will be no less than one report monthly. Reporting by the project Program Manager to the company's corporate management will be in the form of Program Manager's Review (PMR).

Use/Interpretation. Tracking by individual labor category may be done for projects in order to monitor aspects of a particular program that are deemed worthy of special attention. Total personnel are the sum of experienced and support personnel. Special skills personnel are counted within the broad categories of experienced and support, and may be tracked separately:

> *Special Skills Personnel.* Special skills personnel are defined as those individuals who possess specialized software-related abilities defined as crucial to the success of the particular system. For example, Ada programmers are defined as having skills necessary for completing one type of software project but not others.
>
> *Experienced Personnel.* Experienced personnel are defined as degreed individuals with a minimum of three years experience in software development for similar applications.
>
> *Support personnel.* Support personnel are degreed and non-degreed individuals with a minimum of three years experience in software development other than those categorized as experienced software engineers.

Reporting Examples. See Figures C-11 and C-12.

Development Progress Metrics. Progress metrics provide indications of the degree of completeness of the software development effort and can be used to judge readiness to proceed to the next stage of software development. In certain instances, consideration must be given to a possible rebaselining of the software or the addition of modules due to changing requirements.

This metric is used to track the ability to keep Computer Software Unit (CSU) design, code, test, and integration activities on schedule. The development progress metrics should be used with the manpower metrics to identify specific projects that may be having problems.

These metrics should also be used with the Breadth, Depth, and Fault Profile testing metrics to assess the readiness to proceed to a formal test.

Life Cycle Application. Collection is to begin at Software Requirements Review (SRR) and continue for the entirety of the software development. The progress demonstrated in design, coding, unit testing, and integration of CSUs should occur at a reasonable rate. The actual progress shown in these areas versus the originally planned progress can indicate potential problems with a project's schedule.

Algorithm/Graphical Display. The progress metric graphical display should reflect, by project software deliverable item, the percent of planned and actual CSUs per tasking cycle milestone (e.g., % complete/per month). Completed (actual) CSUs may be subcategorized into percent of CSUs 100% designed, percent of CSUs coded and successfully unit tested, and percent of CSUs 100% integrated.

Additionally, using the requirements traceability matrix, the developed and verified functionality versus time may be plotted as a measure of development progress.

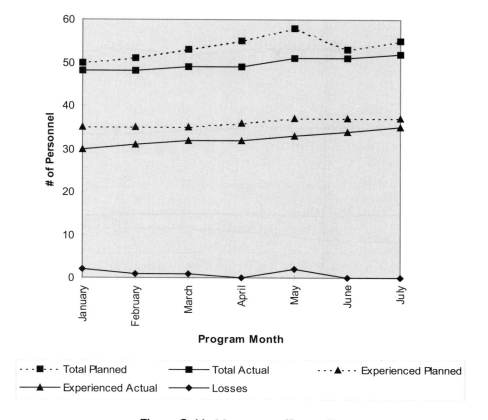

Figure C-11. Manpower staffing profile.

Data Requirements. Project development, test, and integration schedules. For each project project:

1. The number of CSUs
2. Number of CSUs 100% designed
3. Number of CSUs 100% coded and successfully validated
4. Number of CSUs 100% integrated
5. Number of planned CSUs to be 100% designed and reviewed for development cycle

Also include the number of planned CSUs to be 100% coded and successfully validated for development cycle and the number of planned CSUs to be 100% integrated for development cycle.

Frequency and Type of Reporting. The Project Manager shall report progress metric information to the project Program Manager no less than once a month. Also, there shall be no less than one report monthly by the Program Manager to the company's corporate management and customer in the form of Program Manager's Review (PMR).

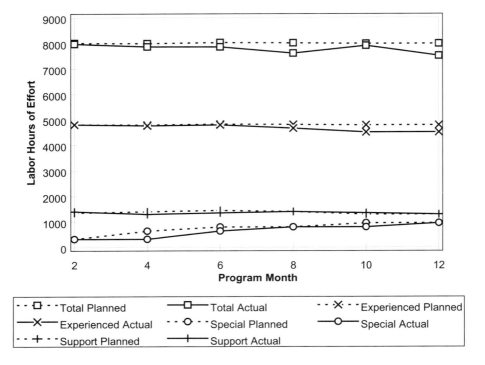

Figure C-12. Manpower metric.

Use/Interpretation. It is important to remember that these metrics pass no judgment on whether the objectives in the development plan can be achieved. Special attention should be paid to the development progress of highly complex CSUs.

Reporting Example. See Figure C-13.

Schedule Metrics. These metrics indicate changes and adherence to the planned schedules for major milestones, activities and key software deliverables. Software activities, and delivery items may be tracked using the schedule metrics. The schedule metrics may be used with other metrics to help judge program risk. For example, it could be used with the test coverage metrics to determine if there is enough time remaining on the current schedule to allow for the completion of all testing.

Life Cycle Application. The Project Manager will begin data collection at project start and continue for the entire software development cycle.

Algorithm/Graphical Display. The Project Manager will plot planned and actual schedules for major milestones and key software deliverables as they change over the tasking cycle. Any milestone event of interest may be plotted.

The schedule metrics may be plotted as they change over time (reflecting milestone movement), providing indications of problems in meeting key events or deliveries. The further to the right of the trend line for each event, the more problems are being encountered.

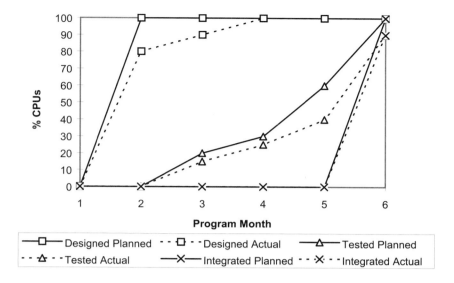

Figure C-13. Development progress metric.

Data Requirements. For all projects, include the following:

1. Software project plan
2. Major milestone schedules
3. Delivery schedule for key software items

For the activity or event tracked during tasking period, include the following:

1. Event name
2. Date of report
3. Planned activity start date
4. Actual activity start date
5. Planned activity end date
6. Actual activity end date

Frequency and Type of Reporting. The Project Manager will report schedule information to the Program Manager and the customer no less than biweekly. This reporting is accomplished as part of scheduled staff meetings. Also, there shall be no less than one report monthly by the Program Manager to the company's corporate management and the customer in the form of a Program Manager's Review (PMR).

Use/Interpretation. No formal evaluation criteria for the trend of the schedule metrics are given. Large slippages are indicative of problems. Maintaining conformance with calendar-driven schedules should not be used as the basis for proceeding beyond milestones. The schedule metric passes no judgment on the achievability of the development plan.

Reporting Example. See Figure C-14.

STATUS

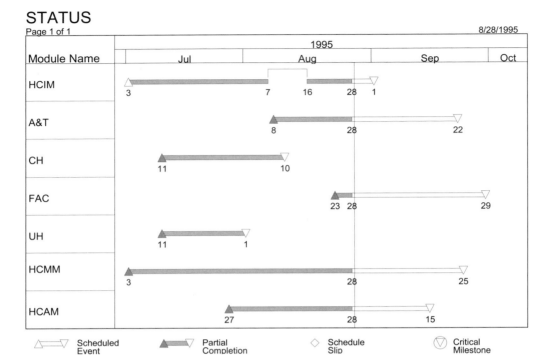

Figure C-14. Example graphical schedule report.

Requirements Category

Requirements Traceability Metric. The requirements traceability metric measures the adherence of the software products to their requirements throughout the development life cycle. The technique for performing this is the development of a project Software Requirements Traceability Matrix (SRTM). (For additional information, refer to the Software Requirements Specification).

The SRTM is the product of a structured, top-down hierarchical analysis that traces the software requirements through the design to the code and test documentation. The requirements traceability metrics should be used in conjunction with the test coverage metrics (depth and breadth of testing) and the development progress metric (optional) to verify if sufficient functionality has been demonstrated to warrant proceeding to the next stage of development or testing. They should also be used in conjunction with the design stability and requirements stability metrics.

Life Cycle Application. The Project Manager will begin tracing metrics during user requirements definition, updating the trace in support of major milestones or at major software release points.

By the nature of the software development process, especially in conjunction with an evolutionary development strategy, the trace of requirements is an iterative process. That is, as new software releases add more functionality to the system, the requirements trace will have to be revisited and augmented.

Algorithm/Graphical Display. This metric is a series of percentages that can be calculated from the matrix described. These will include:

- Percent baseline software requirements specification requirements in CSU design
- Percent software requirements in code
- Percent software requirements having test cases identified

Tracing from code to requirements should also be accomplished. Each CSU is examined for the requirements it satisfies and the percentage of open or additional requirements traced. Backward tracing of requirements at the early stages can identify requirements that have been added in the requirements decomposition process. Each requirement added to the specification should be attributable in some way to a higher-level requirement from a predecessor requirement.

Data Requirements. For all projects, indicate:

1. Documented/Approved Software Requirements Specification (SRS)
2. Documented/Approved Software Design Document (SDD)
3. Software test description in accordance with DOD-STD-498
4. Completed SRTM

For each project software deliverable, indicate:

Number of SRS software requirements:
    Total traceable to SRS
    Traceable to SDD
    Traceable to code
    Traceable to system test

Frequency and Type of Reporting. The Program Manager will update periodically in support of milestones or major releases. This tracing should be a key tool used at all system requirement and design reviews. It can serve to indicate those areas of requirements or software design that have not been sufficiently thought out. The trend of the SRTM should be monitored over time for closure.

Use/Interpretation. A benefit of requirements traceability is that those modules that appear most often in the matrix (thus representing the ones that are most crucial in that they are required for multiple functions or requirements) can be highlighted for earlier development and increased test scrutiny.

Example. An example requirements traceability matrix is shown in Table C-14.

Requirements Stability Metric. Requirements stability metrics indicate the degree to which changes in the software requirements and/or the misunderstanding of requirements implementation affect the development effort. It also allows for determining the cause of requirements changes. The metrics for requirements stability should be used in conjunc-

Table C-14. Requirements Traceability Matrix*

Requirement Name	Priority	Risk	SRS Paragraph	Formal test Paragraph	Test Activity
Input Personnel Data	H	L	3.1.1	6.3.14	D
Store Personnel Data	M	H	3.1.2	6.3.17	T, A

*For additional information regarding requirements elicitation refer to the Software Requirements Management Plan.

Legend: Verification Method(s): D = Demonstration, T = Test, A = Analysis, I = Inspection.

Field Descriptions

1. Requirement Name	A short description of the requirement to be satisfied.
2. Priority	High (H), Medium (M), Low (L); to be negotiated with customer and remains fixed throughout software development lifecycle.
3. Risk	High (H), Medium (M), Low (L); determined by technical staff and will change thoughout software development lifecycle.
4. SRS Paragraph	Paragraph number from Section 3 of the SRS.
5. Validation Method(s)	Method to be used to validate that the requirement has been satisfied.
6. Formal Test Paragraph	This column will provide a linkage between the SRS and the software Test Plan. This will indicate the test to be executed to satisfy the requirement.

tion with those for requirements traceability, fault profiles, and the optional development progress.

Life Cycle Application. Changes in requirements tracking begin during requirements definition and continue to conclusion of the development cycle. It is normal when program development begins that the details of its operation and design be incomplete. It is normal to experience changes in the specifications as the requirements become better defined over time. When design reviews reveal inconsistencies, a discrepancy report is generated. Closure is accomplished by modifying the design or the requirements. When a change is required that increases the scope of the project, a Baseline Change Request (BCR) should be submitted.

Algorithm/Graphical Display. This metric may be presented as a series of percentages representing:

1. The percent of requirements discrepancies, both cumulative and closed, over the life of the project, and cumulative requirements discrepancies over time, versus closure of those discrepancies. Good requirements stability is indicated by a leveling off of the cumulative discrepancies curve, with most discrepancies having reached closure.

2. The percent of requirements changed or added, noncumulative, spanning project development, and the effect of these changes in requirements in lines of code. Several versions of this type of chart are possible. One version may show the number of change enhancement requests (CERs) and affected lines of code. Additionally, it is also possible to look at the number of modules affected by requirements changes.

The plot of open discrepancies can be expected to spike upward at each review and to diminish thereafter as the discrepancies are closed. For each engineering change, the amount of software affected should be reported in order to track the degree to which BCRs increase the difficulty of the development effort. Only those BCRs approved by the associated CCB should be counted.

Data Requirements. A "line of code" will differ from language to language, as well as from programming style to programming style. In order to consistently measure source lines of code, count all noncommented, nonblank, executable and data statements (standard noncommented lines of code or SNCLC).

For each project, indicate:

1. Number of software requirements discrepancies as a result of each review (Software Requirements Review, Preliminary Design Review, and so on)
2. Number of baseline change requests (BCRs) generated from changes in requirements.
3. Total number of SNCLC.
4. Total number of SRS requirements.
5. Percentage of SRS requirements added due to approved BCRs.
6. Percentage of SRS requirements modified due to approved BCRs.
7. Percentage of SRS requirements deleted due to approved BCRs.

Frequency and Type of Reporting. The Project Manager will update the metrics periodically in support of milestones or major releases. This tracing should be a key tool used at all system requirement and design reviews.

Use/Interpretation. Causes of project turbulence can be investigated by looking at requirements stability and design stability together. If design stability is low and requirements stability is high, the designer/coder interface is suspect. If design stability is high and requirements stability is low, the interface between the user and the design activity is suspect. If both design stability and requirements stability are low, both the interfaces between the design activity and the code activity and between the user and the design activity are suspect.

Allowances should be made for higher instability in the case in which rapid prototyping is utilized. At some point in the development effort, the requirements should be firm so that only design and implementation issues will cause further changes to the specification.

A high level of instability at the critical design review (CDR) stage indicates serious problems that must be addressed prior to proceeding to coding.

Example. An example of a graphical display of requirements stability is shown in Figure C-15.

Quality Category

Design Stability Metric. Design stability is used to indicate the quantity of change made to a design. The design progress ratio shows how the completeness of a design advances over time and helps give an indication of how to view the stability in relation to the total projected design. The design stability metric can be used in conjunction with the com-

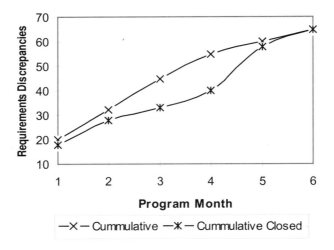

Figure C-15. Requirements stability.

plexity metric to highlight changes to the most complex modules. It can also be used with the requirements metrics to highlight changes to modules that support the most critical user requirements. Design stability does not assess the quality of design. Other metrics (e.g., complexity) can contribute to such an evaluation.

Life Cycle Application. Begin tracking no later than preliminary design review (PDR) and continue for each version until completion.

Algorithm/Graphical Display. Items to be measured are:

M = Number of CSUs in current delivery/design.
F_c = Number of CSUs in current delivery/design that include design.
F_a = Number of CSUs in current delivery/design that are additions to previous delivery.
F_d = Number of CSUs in previous delivery/design that have been deleted.
T = Total modules projected for project.

Formulae:

$$S \text{ (stability)} = [M - (F_a + F_c + F_d)]/M$$

Note: It is possible for stability to be a negative value. This may indicate that everything previously delivered has been changed and that more modules have been added or deleted.

$$DP \text{ (design progress ratio)} = M/T$$

Note: If some modules in the current delivery are to be deleted from the final delivery, it is possible for design progress to be greater than one.

Data Requirements. For each project and version, indicate:

1. Date of completion
2. *M*
3. F_c
4. F_a
5. F_d
6. *T*
7. *S*
8. *DP*

Frequency and Types of Reporting. The Project Manager will update periodically in support of milestones or at major releases. Design stability should be monitored to determine the number and potential impact of design changes, additions, and deletions on the software configuration.

Use/Interpretation. The trend of design stability provides an indication of whether the software design is approaching a stable state, or leveling off, of the curve at a value close to or equal to one. It is important to remember that during periods of inactivity, the magnitude of the change is relatively small or diminishing and may be mistaken for stability.

In addition to a high value and level curve, the following other characteristics of the software should be exhibited:

1. The development progress metric is high.
2. Requirements stability is high.
3. Depth of testing is high.
4. The fault profile curve has leveled off and most software trouble reports [test problem reports (TPRs) or internal test reports (ITRs)] have been fixed or voided.

This metric does not measure the extent or magnitude of the change within a module or assess the quality of the design. Other metrics (e.g., complexity) can contribute to this evaluation.

Allowances should be made for lower stability in the case of rapid prototyping or other development techniques that do not follow the project life cycle model for software development (for additional information, refer to the project's Software Quality Assurance Plan). An upward trend with a high value for both stability and design progress is recommended before release.

Experiences with similar projects should be used as a basis for comparison. Over time, potential thresholds may be developed for similar types of projects.

Reporting Example. A graphical example of a design stability metric is shown in Figure C-16.

Breadth of Testing Metric. Breadth of testing addresses the degree to which required functionality has been successfully demonstrated as well as the amount of testing that has been performed. This testing can be called "black-box" testing, since one is only concerned with obtaining correct outputs as a result of prescribed inputs.

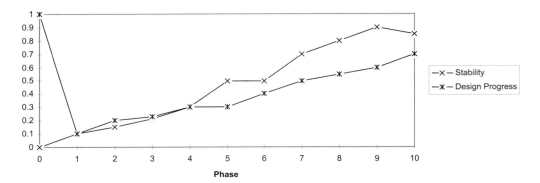

Figure C-16. Design stability/progress metric.

The organization should clearly specify what functionality should be in place at each milestone during the development lifecycle. At each stage of testing (unit through system) emphasis should be placed on demonstrating that a high percentage of the functionality needed for that stage of testing is achieved.

Life Cycle Application. Begin collecting at the end of unit testing. Each priority level of requirements test case to be used for evaluating the success of a requirement should be developed in coordination with project test. This is to ensure that sufficient test cases are generated to adequately demonstrate the requirements.

Algorithm/Graphical Display. Breadth of testing consists of three different measures. One measure deals with coverage and two measures deal with success. These three sub-elements are portrayed in the following equations:

$$\text{Coverage} \quad \times \quad \text{Test Success} \quad = \quad \text{Overall Success}$$

$$\frac{\text{\# CSUs tested}}{\text{total \# CSUs}} \times \frac{\text{\# CSUs passed}}{\text{\#CSUs tested}} = \frac{\text{\# CSUs passed}}{\text{total \# CSUs}}$$

Breadth of Testing Coverage. Breadth of testing coverage is computed by dividing the number of requirements that have been tested (with all applicable test cases under both representative and maximum stress loads) by the total number of requirements.

Breadth of Testing Success. Breadth of testing success is computed by dividing the number of requirements that have been successfully demonstrated through testing by the number of requirements that have been tested.

Breadth of Testing Overall Success. Breadth of testing overall success is computed by dividing the number of requirements that have been successfully demonstrated through testing by the total number of requirements.

Data Requirements. For each project, indicate:

1. Number of SRS requirements and associated priorities.
2. Number of SRS requirements tested with all planned test cases.

3. Number of SRS requirements and associated priorities successfully demonstrated through testing.

4. For each of the four priority levels for additional requirements (SPR, BCR):

   a. Number of additional requirements tested with all planned test cases.

   b. Number of additional requirements demonstrated through testing.

5. For each of the four Alpha/Beta (ITR, TPR) test requirement priority levels:

   a. Number of requirements.

   b. Number of requirements validated with planned test cases.

   c. Number of requirements successfully demonstrated through testing.

6. Test Identification (e.g., unit test, alpha test, beta test)

Frequency and Types of Reporting. All three measures of breadth of testing will be tracked for all project requirements. The results of each measure will be reported at each project level throughout software functional and system level testing at each project milestone and to the company's corporate management monthly.

It is suggested that coverage or success values be expressed as percentages by multiplying each value by 100 before display in order to facilitate understanding and commonality among metrics presentations.

Use/Interpretation. One of the most important measures reflected by this metric is the categorization of requirements in terms of priority levels. With this approach, the most important requirements can be highlighted. Using this prioritization scheme, one can partition the breadth of the testing metric to address each priority level. At various points along the development path, the pivotal requirements for that activity can be addressed in terms of tracing, test coverage, and test success.

Breadth of Testing Coverage. The breadth of testing coverage indicates the amount of testing performed without regard to success. By observing the trend of coverage over time, it is possible to derive the full extent of testing that has been performed.

Breadth of Testing Success. The breadth of testing success provides indications about requirements that have been successfully demonstrated. By observing the trend of the overall success portion of breadth of testing over time, one gets an idea of the growth in successfully demonstrated functionality.

Failing one test case results in a requirement not being fully satisfied. If sufficient resources exist, breadth of testing may be addressed by examining each requirement in terms of the percent of test cases that have been performed and passed. In this way, partial credit for testing a requirement can be shown (assuming multiple test cases exist for a requirement), as opposed to an "all or nothing" approach. This method, which is not mandated, may be useful in introducing additional granularity into breadth of testing.

Example. A graphical example of breadth of testing is shown in Figure C-17.

Depth of Testing Metric. The depth of testing metric provides indications of the extent and success of testing from the point of view of coverage of possible paths/conditions within the software. Depth of testing consists of three separate measures, each of which is

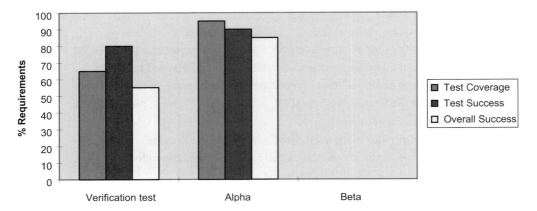

Figure C-17. Breadth of testing.

comprised of one coverage and two success subelements (similar to breadth of testing). The testing can be called "white-box" testing, since there is visibility into the paths/conditions within the software.

This metric should be used in conjunction with requirements traceability and fault profiles and the optional complexity and development progress metrics. They must also be used with breadth of testing metrics to einsure that all aspects of testing are consistent with beta test requirements.

Life Cycle Application. Begin collecting data at critical design review (CDR) and continue through development as changes occur in design, implementation, or testing. Revisit as necessary.

Algorithm/Graphical Display. Graphically, this metric is represented as a composite of the module-level path metrics. This should reflect test coverage and overall success of testing.

Module-Level Metrics. There are four of these:

1. *Path metric.* Defined as the number of paths* in the module that have been successfully executed at least once, divided by the total number of paths in the module.
2. *Statement metric.* The number of executable statements in each module that have been successfully exercised at least once, divided by the total number of executable statements in the module.
3. *Domain metric.* The number of input instances that have been successfully tested with at least one legal entry and one illegal entry in every field of every input parameter, divided by the total number of input instances in the module.
4. *Decision point metric* (optional). The number of decision points in the module that have been successfully exercised with all classes of legal conditions as well as one illegal condition (if any exist) at least once, divided by the total number of decision

*A path is defined as a logical traversal of a module, from an entry point to an exit point.

points in the module. Each decision point containing an "or" should be tested at least once for each of the condition's logical predicates.

Data Requirements. For each module, indicate:

1. Software project name
2. Module identification
3. Number of paths
4. Number of statements
5. Number of input instances
6. Number of paths tested
7. Number of statements tested
8. Number of input instances tested
9. Number of decision points tested
10. Number of paths successfully executed (at least once)
11. Number of statements that have been successfully exercised
12. Number of inputs successfully tested with both legal and illegal entries
13. Number of decision points (optional)
14. Number of decision points that have been successfully exercised at least once with all legal classes of conditions and one illegal condition (optional)

Frequency and Type of Reporting. Report only those modules that have been modified or further tested after depth of testing values have been reported for the first time. In recognition of the effort required to collect and report this metric, the following rules are offered:

1. Always report the domain metric.
2. Compute the path and statement metric if provided with automated tools.

Use/Interpretation. The depth of testing metric addresses the issues of test coverage, test success, and overall success by considering the paths, statements, inputs, and decision points of a software module.

These metrics are to be initially collected at the module level, but this may be extended to the system level if appropriate. Early in the testing process, it makes sense to assess depth of testing at the module level, but later it may make more sense to consider the system in its entirety. This determination is made by the project's Program Manager in conjunction with System Test/Validation.

Reporting Example. A graphical example of depth of testing is shown in Figure C-18.

Fault Profiles Metric. Fault profiles provide insight into the quality of the software, as well as the contractor's ability to fix known faults. These insights come from measuring the lack of quality (i.e., faults) in the software.

Life Cycle Application. Begin after completion of unit testing, when software has been brought under configuration control, and continue through software maintenance.

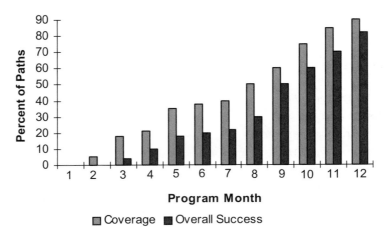

Figure C-18. Depth of testing.

Algorithm/Graphical Display. For each software project plot:

1. The cumulative number of detected and closed software faults (i.e., TPR/ITRs) will be tracked as a function of time. If manpower allows, one plot will be developed for each priority level.
2. The number of software faults detected and software faults closed will be tracked as a function of time.
3. Open faults, tracked by module and priority.
4. Calculate the average age of closed faults as follows. For all closed TPR/ITRs, sum the days from the time the STR was opened and when it was closed. Divide this by the total number of closed problem reports. If manpower allows, this should be calculated for each problem priority or overall.

Data Requirements. The following information can be derived from all change enhancement requests (CERs) that are of the TPR, SPR, or ITR variety:

1. CER unique identifier
2. Date initiated
3. Descriptive title of problem
4. Detailed description of problem
5. Priority:
   a. 1 = causes mission-essential function to be disabled or jeopardizes personnel safety. No workaround.
   b. 2 = causes mission essential function to be degraded. There is a reasonable workaround.
   c. 3 = causes operator inconvenience but does not affect a mission-essential function.
   d. 0 = all other errors.

6. Category:
    a. Requirements = Baseline Change Request (BCR)
    b. Alpha Test = Internal Test Report (ITR)
    c. Beta Test = Test Problem Report (TPR)
    d. User-driven changes to data = Software Problem Report (SPR)
    e. User-driven change to interfaces = Interface Change Notice (ICN).
7. Status:
    a. Open
    b. Voided
    c. Hold
    d. Testing (Alpha)
    e. Fixed
8. Date detected.
9. Date closed.
10. Software project, module, and version.
11. Estimated required effort and actual effort required.
12. The following reports may be generated using the SCRTS reporting capabilities. These items are the building block for graphical representations and fault profiles. Indicate for each module by priority:
    a. Cumulative number of CERs
    b. Cumulative number of closed CERs
    c. Average age of closed CERs
    d. Average age of open CERs
    e. Average age of CERs (both open and closed)
    f. Totals for each CER category (described above)

Frequency and Type of Reporting. The Project Manager will update periodically in support of milestones or at major releases. Design stability should be monitored to determine the number and potential impact of design changes, additions, and deletions on the software configuration. The types of items examined are:

1. Displays of detected faults versus closed (verified) faults should be examined for each problem's priority level and for each module. Applied during the early stages of development, fault profiles measure the quality of the translation of the software requirements into the design. CERs opened during this time may suggest that requirements are either not being defined or are being interpreted correctly. Applied later in the development process, assuming adequate testing, fault profiles measure the implementation of requirements and design into code. CERs opened during this stage could be the result of having an inadequate design to implement those requirements, or a poor implementation of the design into code. An examination of the fault category should provide indications of these causal relationships.

2. If the cumulative number of closed ITR/TPRs remains constant over time and a number of them remain open, this may indicate a lack of problem resolution. The

age of the open STRs should be checked to see if they have been open for an unreasonable period of time. If so, these areas need to be reported to the project's Program Manager as areas of increased risk.

3. The monthly noncumulative ITR totals for each project can be compared to the TPR totals to provide insights into the adequacy of the Alpha validation program.

4. Open-age histograms can be used to indicate which modules are the most troublesome with respect to fixing faults. This may serve to indicate that the project team may need assistance.

5. Average-age graphs can track whether the time to close faults is increasing with time, which may be an indication that the developer is becoming saturated or that some faults are exceedingly difficult to fix. Special consideration should be paid to the gap between open and fixed CERs. If a constant gap or a continuing divergence is observed, especially as a major milestone is approached, the Program Manager should take appropriate action.

Once an average CER age has been established, large individual deviations should be investigated.

Use/Interpretation. Early in the development process, fault profiles can be used to measure the quality of the translation of the software requirements into the design. Later in the process, they can be used to measure the quality of the implementation of the software requirements and design into code.

If fault profile tracking starts early in software development, an average CER open age of less than three months may be experienced. After release, this value may be expected to rise.

Reporting Examples. A graphical example of cumulative reporting by month is shown in Figure C-19. Figure C-20 shows reporting of project CERs by month. Figure C-21 shows reporting of average age versus open faults.

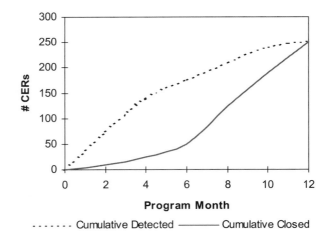

Figure C-19. Cumulative reporting by month.

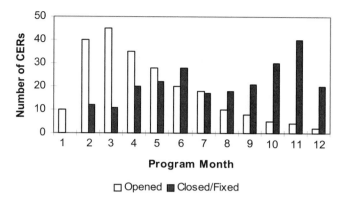

Figure C-20. Reporting of project CERs by month.

Measurement Information Model in ISO/IEC 15939

Table C-15 provides three examples of instantiations of the Measurement Information Model as presented in ISO/IEC 15939 Software Engineering—Software Measurement Process [85].

REQUIREMENTS DEVELOPMENT

Change Enhancement Requests

The identification and tracking of application requirements is accomplished through some type of change enhancement request (CER). A CER is used to control changes to all documents and software under SCM control and for documents and software that have been released to SCM. Ideally, the CER is an on-line form that the software developers use to submit changes to software and documents to the SCM Lead for updating the product. Table C-16 provides an example of the types of data that may be captured and tracked in support of application development or modification.

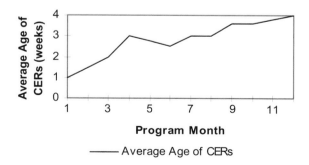

Figure C-21. Average age of CERs by project.

Table C-15. Measurement information model in ISO/IEC 15939

	Productivity	Quality	Progress
Information need	Estimate productivity of future project	Evaluate product quality during design	Assess status of coding activity
Measurable concept	Project productivity	Product quality	Activity status
Relevant entities	1. Code produced by past projects 2. Effort expended by past projects	1. Design packages 2. Design inspection reports	1. Plan/schedule 2. Code units completed or in progress
Attributes	1. C++ language statements (in code) 2. Timecard entries (recording effort)	1. Text of inspection packages 2. Lists of defects found in inspections	1. Code units identified in plan 2. Code unit status
Base measures	1. Project X lines of code 2. Project X hours of effort	1. Package X size 2. Total defects for package X	1. Code units planned to date 2. Code unit status
Measurement method	1. Count semicolons in Project X code 2. Add timecard entries together for Project X	1. Count number of lines of text for each package 2. Count number of defects listed in each report	1. Count number of code units scheduled to be completed by this date 2. Ask programmer for percent complete of each unit
Type of measurement method	1. Objective 2. Objective	1. Objective 2. Objective	1. Objective 2. Subjective
Scale	1. Integers from zero to infinity 2. Real numbers from zero to infinity	1. Integers from zero to infinity 2. Integers from zero to infinity	1 Integers from zero to infinity 2. Integers from zero to one hundred
Type of Scale	1. Ratio 2. Ratio	1. Ratio 2. Ratio	1. Ratio 2. Ordinal
Unit of measurement	1. Line 2. Hour	1. Lines 2. Defects	1. Code unit 2. One-hundredth code unit
Derived measure	Project X productivity	Inspection defect density	Progress to date
Measurement function	Divide Project X lines of Code by Project X hours of Effort	Divide total defects by package size for each package	Add status for all code units planned to be complete to date
Indicator	Average productivity	Design defect density	Code status expressed as a ratio
Model	Compute mean and standard deviation of all project productivity values	Compute process center and control limits using values of defect density	Divide progress to date by (code units planned to date times 100)

Table C-15. *Continued*

	Productivity	Quality	Progress
Decision criteria	Computed confidence limits based on the standard deviation indicate the likelihood that an actual result close to the average productivity will be achieved. Very wide confidence limits suggest a potentially large departure and the need for contingency planning to deal with this outcome.	Results outside the control limits require further investigations.	Resulting ratio should fall between 0.9 and 1.1 to conclude that the project is on schedule.

Baseline Change Request

The tracking of changes to the development baseline may be accomplished using a baseline change request (BCR). A BCR is a variation of the basic CER that is used to control changes to all documents and software under SCM control and for documents and software that have been released to SCM. Ideally, the BCR is an on-line form that the software developers use to submit changes to software and documents to the SCM Lead for updating the product. Figure C-22 provides an example of the types of data that may be captured and tracked in support of application development or modification. *Baseline Change Request.doc* provides an electronic version of this BCR on the CD-ROM accompanying this book.

TECHNICAL SOLUTION

Work-Breakdown Structure (WBS) for Postdevelopment Stage

The example provided in Table C-17 provides suggested content for a WBS in support of the postdevelopment stage of a software development lifecycle. This WBS content is illustrative only and should be customized to meet specific project requirements.

Architecture Design Success Factors and Pitfalls [72]

Through a series of software architecture workshops, Bredemeyer Consulting (www.bredemeyer.com) has identified what they call the top critical success factors for an architecting effort. These are identified in Table C-18.

Conversely, the top pitfalls associated with an architecting effort have also been identified. These are shown in Table C-19.

UML Modeling

Figure C-23 uses the UML notation to illustrate the inheritance, aggregation, and reference relationships between work product sets, their work products, their versions, the languages they are documented in, and the producers that produce them.

Table C-16. Typical elements in a CER

Element	Values
CER #	Unique CER identifier assigned by CRC.
Type	One of the following: BCR—baseline change request; CER indicating additional/changed requirement SPR—software problem report; CER indicating software problem identified externally during or after beta test ITR—internal test report; CER indicating software problem identified internally through validation procedures DOC—documentation change; CER indicting change to software documentation
Status	One of the following: Open—CER in queue for work assignment. Testing—CER in validation test Working—CER assignment for implementation as SCR(s) Voided—CER deemed not appropriate for software release Hold—CER status to be determined Fixed—CER incorporation into software complete
Category	One of the following: Data—Problem resulting from inaccurate data processing Doc—Documentation inaccurate Reqt—Problem caused by inaccurate requirement
Priority	One of the following: 1. Highest priority; indicates software crash with no workaround 2. Cannot perform required functionality; available workaround 3. Lowest priority; not critical to software performance 0. CER has no established priority
Date Submitted	The date CER initiated; defaults to current date
Date Closed	The date CER passes module test
Originator	Identifies the source of CER
Problem Description	Complete and detailed description of the problem, enhancement, or requirement
Short Title	Short title summarizing CER
SRS Ref #	Cross-reference to associated software requirements specification (SRS) paragraph identifier
SDD Ref #	Cross-reference to associated software design document (SDD) paragraph identifier(s)
STP Ref #	Cross-reference to associated software test plan (STP) paragraph identifier(s)
Targeted Version	Software module descriptor and version number; determined by SCM Lead
File	Files affected by CER
Functions Affected	List of functions affected by CER
Time Estimated	Estimated hours until implementation complete
Time Actual	Actual hours to complete CER implementation
Module	Name of module affected by CER

Table C-16. *Continued*

Element	Values
Problem Description	Description of original CER, add comments regarding implementation
Release	Release(s) affected by CER
CER #	Cross-reference number used to associate item with other relevant CERs
Software Engineering Name	Last name of engineer implementing CER

Originator Module:		Priority:
Name:	Title:	BCR No.:
Organization:	SRS Revision:	Date:
BCR Title:		
Change Summary:		
Reason for Change:		
Impact if Change Not Made:		

Org	Name Yes		No	Desired Applicability
				Release Available:
				Release Unsupported:
				CCB Approval
				Name:
				Date:
				Signature:
				_____ Approved
				_____ Disapproved
				Comments:

Figure C-22. Baseline change request.

Table C-17. WBS content for postdevelopment activity

Product Installation
   Distribute Product (software)
      Package and distribute product
      Distribute installation information
      Conduct integration test for product
      Conduct regression test for product
      Conduct user acceptance test for software
      Conduct reviews for product
      Perform configuration control for product
      Implement documentation for product
   Install Product (software)
      Install product (packaged software) and database data
      Install any related hardware for the product
      Document installation problems
   Accept Product (software) in Operations Environment
      Compare installed product (software) to acceptance criteria
      Conduct reviews for installed product (software)
      Perform configuration control for installed product (software)

Operations and Support
   Utilize installed software system
   Monitor performance
   Identify anomalies
   Produce operations log
   Conduct reviews for operations logs
   Perform configuration control for operations logs

Provide Technical Assistance and Consulting
   Provide response to technical questions or problems
   Log problems

Maintain Support Request Logs
   Record support requests
   Record anomalies
   Conduct reviews for support request logs

Maintenance
   Identify Product (software) Improvements Needs
      Identify product improvements
      Develop corrective/perfective strategies
      Produce product (software) improvement recommendations
   Implement Problem Reporting Method
      Analyze reported problems
      Produce report log
      Produce enhancement problem reported information
      Produce corrective problem reported information
      Perform configuration control for reported information
   Reapply Software Life Cycle methodology

Table C-18. Software architecting success factors

1. The strategic business objective(s) of key sponsor must be addressed and the lead architect and team must have a sound understanding of functional requirements. There must be a good match between the technology and the business strategy.
2. The project must have a good lead architect with a well-defined role who:
   Exhibits good domain knowledge
   Is a good communicator/listener
   Is competent in project management
   Has a clear and compelling vision
   Is able to champion the cause
3. The architecting effort must contribute immediate value to the developers. The architecture must be based upon clear specifications.
4. There should be architectural advocates at all levels of the organization, ensuring that there is strong management sponsorship and that required talent and resources are available.
5. Architecture must be woven into the culture to the point that the validation of requirements occurs during each step of the process.
6. The customer must be involved and his expectations clearly defined.
7. Strong interpersonal and team communication and ownership must be present.

Make/Buy Decision Matrix

A make/buy decision matrix may be used to support the CMMI specific practice in support of Technical Solution, SP 2.4—Perform Make, Buy, or Reuse Analyses. According to the SEI:

> Software enters an organization in one of three ways: it can be *built* in-house, *purchased* from a commercial vendor, or *commissioned* through a third party to be built especially for the organization. Software that is built in-house can actually be constructed anew or *mined* from software already in the organization for use in a new effort. Every piece of software that is part of a development effort arrived as the result of this unavoidable four-way choice, which we dub "make/buy/mine/commission." Organizations that build software systems all make this choice—it cannot be avoided—but almost always without a conscious rationale for the alternative they select. The purpose of this practice area is both to underscore the necessity of making a conscious and reasoned choice and to describe some of the analyses that are appropriate for helping to make this choice. [58]

Table C-19. Software architecting pitfalls

1. The lead architect, or supporting senior management, does not effectively lead the effort. Roles and responsibilities are inadequately defined.
2. Thinking is at too low a level. The "big picture," or integration requirements, are not considered.
3. There is poor communication inside/outside the architecture team.
4. There is inadequate support, including inadequate resources and/or talent.
5. There is a lack of customer focus. The architecture team does not understand or respond to functional requirements.
6. The requirements are unclear, not well defined, not signed off on, or changing.

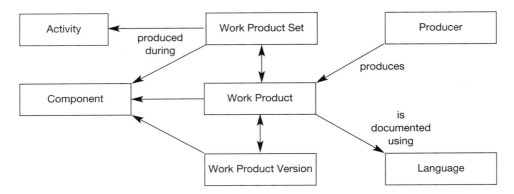

Figure C-23. UML notation.

This matrix may be used to support the analysis of alternative solutions. This matrix should include a listing of all project requirements and their estimated cost. It should also include a list of all candidate applications that come closest to meeting the specifications as well as a list of reusable components. This matrix should be used to compare and analyze key functions and costs, addressing the qualitative as well as quantitative issues surrounding the selection process. An example decision matrix is provided on the CD-ROM accompanying this book and is entitled *Decision Matrix.xls*.

The SEI provides some sample questions to support the analysis of each of the make/buy/mine/commission options that are shown in Table C-20.

Alternative Solution Screening Criteria Matrix

An alternative solution screening criteria matrix may be used to support the analysis of alternative solutions. This is in support of the specific CMMI® practice in support of Technical Solution, SP1.1—Develop Detailed Alternative Solutions and Selection Criteria. This matrix should include a listing of all project requirements and their estimated cost. It should also include a list of all candidate applications that come closest to meeting the specifications as well as a list of reusable components. This matrix should be used to compare and analyze key functions and costs, addressing the qualitative as well as quantitative issues surrounding the selection process. An example decision matrix is provided on the CD-ROM accompanying this book and is entitled *Decision Matrix.xls*. Suggestions in support of effective screening practices are provided in Table C-21.

Unit Test Report

This Unit Test Report template is designed to facilitate the definition of processes and procedures relating to unit test activities. This template was developed using IEEE Std 829-1998, IEEE Standard for Software Test Documentation, and IEEE Std 1008-1987, Software Unit Testing, which has been adapted to support CMMI requirements. The intent is to support compliance to CMMI Technical Solution SP 3.1, Implement the Designs of the Product Components.

Information displayed in blocked text under each section contains suggested items for inclusion in the Unit Test Report. It is important to note that significant additional information is available from IEEE Std 829-1998 and IEEE Std 1008-1987, as the standards

Table C-20. Questions in support of the make/buy/mine/commission decision [58]

Sample questions for the "make" option:
- Are developers with appropriate expertise available?
- How would the developers be utilized if not on this project?
- If additional personnel need to be hired, will they be available within the needed time frame?
- How successful has the organization been in developing similar assets?
- What specific flexibilities are gained by developing products in-house as opposed to purchasing them?
- What development tools and environments are available? Are they suitable? How skilled in their use is the targeted work force?
- What are the costs of development tools and training, if needed?
- What are the other specific costs of developing the asset in-house?
- What are the other specific benefits of developing the asset in-house?

Sample questions for the "buy" option:
- What assets are commercially available?
- How well does the COTS software conform to the product line architecture?
- How closely does the available COTS software satisfy the product line requirements?
- Are small changes in COTS software a viable option?
- Is source code available with the software? What documentation comes with the software?
- What are the integration challenges?
- What rights to redistribute are purchased with the COTS software?
- What are the other specific costs associated with purchasing the software?
- What are the other specific benefits associated with purchasing the software?
- How stable is the vendor?
- How often are upgrades produced? How relevant are the upgrades to the product line?
- How responsive is the vendor to user requests for improvements?
- How strong is the vendor support?

Sample questions for the "mine" option:
- What legacy systems are available from which to mine assets?
- How close is the functionality of the legacy software to the functionality that is required?
- What is the defect track record for the software and nonsoftware assets?
- How well is the legacy system documented?
- What mining strategies are appropriate?
- How expensive are those strategies?
- What experience does the organization have in mining assets?
- What mining tools are available? Are they appropriate? How skilled is the work force in their use?
- What are the costs of mining tools and training, if needed?
- What are the other specific costs associated with mining the asset?
- What are the other specific benefits associated with mining the asset?
- What changes need to be made to the legacy asset to perform the mining?
- What are the costs and risks of those changes?
- What types of noncode assets are available?
- What modifications or additions need to be made to them?

Sample questions for the "commission" option:
- What contractors are available to develop the asset?
- What is the track record of the contractor in terms of schedule and budget?
- Is the acquiring organization skilled in supervising contracted work?
- Are the requirements defined to the extent that the asset can be subcontracted?
- Are interface specifications well defined and stable?
- What experience does the contractor have with the principles of product line development?
- Who needs to own the asset? Who maintains it?

(continued)

Table C-20. Questions in support of the make/buy/mine/commission decision [58]

Sample questions for the "commission" option (*cont.*):
- What are the other specific costs associated with commissioning the asset?
- What are the other specific benefits associated with commissioning the asset?
- What are the costs of maintaining the commissioned asset?
- Does commissioning the asset involve divulging to the contractor any technology or information that it is in the acquiring organization's interest to keep in-house?

---

provide additional tutorial information. Table C-22 provides a sample document outline. The CD-ROM accompanying this book also provides an electronic template, Unit Test Report.doc, in support of this document.

The following provides section-by-section guidance in support of the creation of a Unit Test Report. This guidance should be used to help define a unit test process and should reflect the actual processes and procedures of the implementing organization. Additional information is provided in the document template, *Unit Test Report.doc,* located on the CD-ROM accompanying this book.

Table C-21. Tips for effective solution screening [63]

Strike a reasoned balance between efficiency and completeness	Efficient screening often requires employing high-level criteria that can quickly exclude many COTS components. For example, a common and often appropriate screening strategy suggests that examining only the top few (e.g., 3–5) competitive components in terms of sales or customer-installed base. However, there are often reasons to additionally consider smaller, niche players that provide a unique or tailored capability more in line with system expectations. This suggests that there may be multiple pathways for inclusion of a component.
Avoid premature focus on detailed architecture and design	Although high-level architecture is an appropriate screening criteria, there is a tendency to prematurely focus on detailed architecture and design as screening criteria (for example, specifying the architectural characteristics unique to one component). Where architectural or design decisions have already been made that must be reflected by the chosen component, it is important that criteria reflect these decisions. However, to fight against the tendency to think within the box, consider the risks and potential work-arounds if a highly similar capability is delivered in a different manner.
Optimize the order in which you apply screening criteria	Criteria will vary along dimensions of ability to discriminate and effort required to obtain data. By their nature, some criteria will be highly discriminatory in determining which components are appropriate. Other criteria, though important, will eliminate fewer components from consideration. Still other criteria will require more work than the norm to accumulate data. Consider both how discriminating a criterion is and the effort to evaluate components against the criterion when determining an order for applying screening criteria.
Market share/growth are good screens for long-lived systems	A defining characteristic of many systems is expected lifespan. Often, systems are used for over 20 years. COTS components in these systems should come from financially sound companies with good prospects to remain in business.

Table C-22. Unit Test Report document outline

---

1. Introduction
2. Identification and Purpose
3. Approach/Strategy
4. Overview
5. Test Configuration/Environmental
    5.1   Software Environment
    5.2   Test Suite
    5.3   Test Stub
    5.4   Test Driver
    5.5   Test Harness
    5.6   Preservation of Testing Products
6. Test Cases/Items
    6.1   Test Case Identifier
    6.2   Test Sequence for Interdependencies
    6.3   Items Tested, Indicating their Version/Revision Level
    6.4   Specific Test Data
    6.5   Test Conditions Met
    6.6   Test Procedure/Script/Task Identifier User
    6.7   Test Tool Used
    6.8   Results/Outputs/Summary of Faults
7. Evaluation
    7.1   Termination Requirements Met
    7.2   Test Effectiveness
    7.3   Variances from Requirement # and/or Design #
    7.4   Risks and Contingencies
8. Summary of Activities
9. Notes
10. Approvals

Introduction. This section should provide a brief introductory overview of the project and related testing activities described in this document.

Identification and Purpose. This section should provide information that uniquely identifies all test cases, items, and component descriptions involved in the testing effort. This can also be provided in the form of a unique project plan unit test identifier. The following text provides an example:

> The responsible tester was [Tester Name], [responsible organization], Role [coder, designer, test team member], and unit test completed on [Date/time of test conclusion].
>
> This Unit Test Report # is used to detail the unit testing conducted for the test items/Component Description in [Project Name] ([Project Abbreviation]) Version [xx], Statement of Work, [date], Task Order [to number], Contract No. [Contract #], and Amendments.
>
> The goal of [project acronym] development is to [goal]. This development will allow [purpose].

Approach/strategy. This unit test is the first or bottom level of all tests planned according to the development life cycle. The unit test consists of functional testing, which is testing of the application component(s) against a subset of the operational requirements. The unit test is classified as white-box testing, whose goal is to ensure that code is constructed properly and does not contain any hidden weaknesses.

This section should describe the unit test approach. Identify the risk areas tested. Specify constraints on characteristic determination, test design, or test implementation. Identify existing sources of input, output, and state data. Identify general techniques for data validation. Identify general techniques to be used for output recording, collection, reduction, and validation. Describe provisions for application software that directly interface with the units to be tested.

Overview. This section should provide a summary of all test items/components tested. The need for each item and its history may be addressed here as well. References to the associated project documents should be cited here. The following provides an example:

[The list of individual program units (subprogram, object, package, or module) identifiers and their version/revision levels and Configuration Management Item(s) # are:

Subprogram079	v15	Config12345
Subprogram081	v10	Config12346]

Specifics regarding the testing of these modules are identified in the:
Any specific exclusions should be discussed.
This document describes the Unit Test Report (UTR) for the [PROJECT ABBREVIATION] System.
[PROJECT ABBREVIATION] documents related to this UTR are:

- Project Plan Milestone/Task #
- Software Requirements Traceability Matrix (RTM) line item descriptions:
  - Software Requirements Specification (SRS)
  - Software Development Plan (SDP)
  - Software Design Document (SDD)
- User's Manual/System Administrator's Manual
- Data Dictionary
- Test Problem Report (TPR) #
- Change Enhancement Request (CER) #

This UTR describes the deliverables to be generated in the testing of the [PROJECT ABBREVIATION]. This report complies with the procedures defined in the [PROJECT ABBREVIATION] Software Configuration Management (SCM) Plan and the [PROJECT ABBREVIATION] Software Quality Assurance (SQA) Plan. Any nonconformance to these plans will be documented as such in this UTR.

This document is based on the [PROJECT ABBREVIATION] Unit Test Report template with tailoring appropriate to the processes associated with the creation of the [PROJECT ABBREVIATION]. The information contained in this UTR has been created for [Customer Name] and is to be considered "For Official Use Only."

Test Configuration/Environment. Provide a high-level summary of Test Configuration/ Environment. Depending on system complexity, this could include:

- Software Environment
- Test Suite
- Test Stub
- Test Driver

- Test Harness
- Preservation of Testing Products

Software Environment. This section should provide a complete list of all communications and system software; operating system version; physical characteristics of the facilities, including the hardware; and database version. System and software items used in support of testing should include versions. If the system is kept in an online repository, reference to the storage location may be cited instead of listing all items here. The following are examples:

> System and software used in the testing of [PROJECT ABBREVIATION]: [System and software List].
>
> For details of client and server System specifications see the [PROJECT ABBREVIATION] System Requirements Specifications (SysRS).
>
> Unit tests were to be performed at [company name], [site location], in the test system [test system number], using OS X, 10.3.2.

Test Suite. Describe the set of test cases that serve a particular testing goal.

Test Stub. Describe the temporary, minimal implementation of a component to increase controllability and observability in testing.

Test Driver. Describe the class or utility program that applies test cases to the component.

Test Harness. Describe the system of test drivers and other tools to support test execution.

Preservation of Testing Products. Describe the location of resulting test data, any revised test specifications, any and revised test data.

Test Cases/Items. This section should describe each of the Test Cases/Items. Depending on system complexity, this could include:

- Test case identifier
- Test sequence for interdependencies
- Items tested, indicating their version/revision level
- Specific test data
- Test conditions met
- Test procedure/script/task identifier used
- Test tool used
- Results/outputs summary of faults

Test Case Identifier. This subsection should describe the unique identifier of the test case, allowing others to find and repeat the same test case. The following is provided as an example:

Test Case 051

Test Sequence for Interdependencies. This subsection should describe where the sequence of this test case fits in the hieracrchy or architecture of other test cases identifiers. The following is provided as an example:

Test Case 052 was the 2nd of 8 test cases following Test Cases 04x and preceeding Test Cases 06x

Items Tested, Indicating their Version/Revision Level. This subsection should describe the test items/component tested by this test case. The following is provided as an example:

Subprogram079   v15   Config12345
Subprogram081   v10   Config12346]

Specific Test Data. This subsection should describe the input data and states used by this test case (e.g., queries, transactions, batch files, references, software data structures, or invalid inputs). The following is provided as an example:

Subprogram081   testfile123456

Test Conditions Met. This subsection should describe the resulting test condition for the test case, in which the test condition compares the values of two formulae. If the comparison is true, then the condition is true. Otherwise, it is false. Some typical conditions may be:

- Check for correct handling of erroneous inputs
- Check for maximum capacity
- User interaction behavior consistency
- Retrieving data
- Saving data
- Display screen and printing format consistency
- Measure time of reaction to user input

The following is provided as an example:

Subprogram081   True, created 256 byte testfile123457

Test Procedure/Script/Task Identifier Used. This subsection should describe the tasks conducted for this test case. Also note if there are any constraints on the test procedures, or special procedural requirements, or unusual resources used. The following is provided as an example:

Task062   initiate Subprogram079, followed by Subprogram081

Test Tool Used. This subsection should describe the test case tool use from the test environment. The following is provided as an example:

Test Suite Alpha, Test Driver A1 of Test Harness A

Results/Outputs/Summary of Faults. This subsection should describe the results of using this Test Case. For each failure, have the failure analyzed and record the fault information:

- Case 1: Test Specification or Test Data
- Case 2: Test Procedure Executio
- Case 3: Test Environment
- Case 4: Unit Implementation
- Case 5: Unit Design

The following is provided as an example:

```
Task062              revised to v8 (Case 2)
Testfile123456       revised to v3 (Case 1)
Subprogram081        revised to v8, v9, & v10 (Case 4)]
```

Evaluation. This section includes the following two subsections.

Termination Requirements Met. This subsection should provide a list of the termination requirements met by the unit testing, starting with a summary the evaluation of the test cases/items.

Test Effectiveness. This subsection should identify the areas (for example, features, procedures, states, functions, data characteristics, instructions) covered by the unit test set and the degree of coverage required for each area. Coverage can be stated as a percentage (1 to 100%) for the major objective units:

- Code
- Instructions
- Branches
- Path
- Predicate
- Boundary value
- Features
- Procedures
- States
- Functions

Variances from Requirement # and/or Design #. This section should report any variances of the test items from their design and/or requirements specifications. If this occurred, specify the:

- Reason for each variance
- Identification of defects that are not efficiently identified during previous peer reviews
- Resulting test problem report (TPR) numbers

Risks and Contingencies. This section should identify any unresolved incidents and summary the test item limitations.

Summary of Activities. This section should provide actual measurements for calculation of both test efficiency and test productivity ratios:

- Hours to complete testing
  - Not counting test suspension
- Hours for setup
- Hours test environment used
- Size of testing scope

Note. This section should provide any additional information not previously covered in the unit test report.

Approvals. This section should include the approvals specified in the Project Plan.

PRODUCT INTEGRATION

System Integration Test Report

This System Integration Test Report template is designed to facilitate the definition of processes and procedures relating to system integration test activities. This template was developed using IEEE Std 829-1998, IEEE Standard for Software Test Documentation, and IEEE Std 1008-1987, Software Unit Testing, which has been adapted to support CMMI requirements. The intent is to support Compliance to CMMI Product Integration SP 3.3, Evaluate Assembled Product Components.

Information displayed in blocked text under each section is suggested items for inclusion in the System Integration Test Report. It is important to note that significant additional information is available from IEEE Std 829-1998 and IEEE Std 1008-1987, as the standards provide additional tutorial information. Table C-23 provides a sample docu-

Table C-23. System Integration
Test Report document outline

1. Introduction
2. Identification & Purpose
3. Approach/Strategy
4. Overview
5. Test Configuration
    5.1. Test Environment
    5.2. Test Identification
    5.3. Results/ Outputs/ Summary of Faults
6. Test Schedules
7. Risk Management/Requirements Traceability
8. Notes
9. Approvals

ment outline. An electronic version of this document, called *System Integration Test Report.doc,* is provided on the CD-ROM accompanying this book.

Introduction. This section should provide a brief introductory overview of the project and related testing activities described in the document.

Identification and Purpose. This section should provide information that uniquely identifies the system and the associated testing effort. This can also be provided in the form of a unique project plan milestone identifier. The following text provides an example:

> The responsible lead tester was [Tester Name], [responsible organization], Role [lead test team member], and System Integration Test completed on [Date/time of System Integration Test conclusion], as planned in the System Integration Test Plan.
>
> This System Integration Test Report is to detail the System Integration testing conducted for the [Project Name] ([Project Abbreviation]) Version [xx], Statement of Work, [date], Task Order [to number], Contract No. [Contract #], and Amendments.
>
> The goal of [project acronym] development is to [goal]. This System will allow [purpose].
>
> Specifics regarding the implementation of these system modules are identified in the [Project Abbreviation] System Requirements Specification (SysRS) with line item descriptions in the accompanying Requirements Traceability Matrix (RTM) and references to the Interface Control Documentation (ICD) and the System Integration Test Plan (SysITP).

Approach/Strategy. This System Integration Test follows unit testing of all modules and components scheduled for integration, as planned according to the System Integration Test Plan. The result is the creation of either a partial or complete system. These tests result in discovery of problems that arise from module/component interactions and these problems are localized to specific components for correction and retesting. The System Integration Test includes white-box testing, whose goal is to ensure that modules and components are constructed properly and do not contain any hidden weaknesses.

This section should summarize the deviations from the System Integration Test Plan objectives, kinds (software, databases, hardware, COTS, and prototype usability), and approaches (e.g. top-down, bottom-up, object-oriented, and/or interface testing).

Overview. This section should provide a summary of all system modules and components and system features tested. Any specific exclusion should be discussed. The following provides an example:

> The list of individual program units (subprogram, object, package, or module) identifiers and their version/revision levels and Configuration Management Item(s) # are:
>
> | | | |
> |---|---|---|
> | Subprogram078 | v19 | Config12343 |
> | Subprogram079 | v15 | Config12345 |
> | Subprogram080 | v13 | Config12344 |
> | Subprogram081 | v10 | Config12346 |

This document describes the System Integration Test Report (SysITR) for the [PROJECT ABBREVIATION] System. [PROJECT ABBREVIATION] documentation related to this SysITR is:

> Project Plan Milestone/Task #
> SystemRequirements Specification (SRS),

System Requirements Traceability Matrix (RTM)
Test Problem Report (TPR) #
Change Enhancement Request (CER) #

This SysITR describes the deliverables to be generated in the system integration testing of the [PROJECT ABBREVIATION]. This report complies with the procedures defined in the [PROJECT ABBREVIATION] Software Configuration Management (SCM) Plan and the [PROJECT ABBREVIATION] Software Quality Assurance (SQA) Plan. Any nonconformance to these plans will be documented as such in this SysITR.

This document is based on the [PROJECT ABBREVIATION] System Integration Test Report template with tailoring appropriate to the processes associated with the creation of the [PROJECT ABBREVIATION]. The information contained in this SysITR has been created for [Customer Name] and is to be considered "For Official Use Only."

Test Configuration/Environment. This section should summarize the deviations from the System Integration Test Plan—System Integration Test Environmental. The following is an example:

The System Integration Tests were performed at [company name], [site location]. All integration testing were conducted in the development center [room number]. The deviations from the System Integration Test Plan for the System and software used in the testing of [PROJECT ABBREVIATION] are asterisked below: [System and software List]

Several errors in Test Suite Alpha, Test Driver A1 of Test Harness A were discovered and corrected, resulting in retesting at both unit and subsystem levels and a one day longer test interval, but no additional risk.

Test Identification. This section should summarize the deviations from the System Integration Test Plan—Test Identification. The following is an example:

Subsystem A integration was delayed due to late arrival of Module A12345. This resulted in resequencing Substem B integration testing. A one week longer test interval, but no additional risk, resulted.

Results/Outputs/Summary of Faults. This section should describe the results of system integration testing. The following is an example:

For the 35 failures analyzed and localized to specific components or the test environment, the fault information was:

- Case 1: Test Specification or Test Data
- Case 2: Test Procedure Execution
- Case 3: Test Environment
- Case 4: Component Implementation
- Case 5: Component Design

The seven above Components with defects and corrected version #s were:

Subprogram078	v19	Config12343
Subprogramxxx	vxx	Configxxxxx
. . .		

According to the System Integration Test Plan measurement objectives, 85% of all test scripts passed on the first pass.

Test Schedules. This section should summarize the deviations from the System Integration Test Plan—Test Schedule.

Risk Management/Requirements Traceability. This section should identify all high-risk encountered and the action taken. The following provides an example:

> While all components have been tested and integrated, with only a two weeks extended test interval, several changes are needed in the System Test Plan to fully complete the revised Requirements Traceability matrix.

Summary of Activities. This section should provide actual measurements for calculation of both Test efficiency and Test Productivity ratios:

- Hours to complete testing
  - ○ Not counting test suspension
- Hours for setup
- Hours test environment used
- Size of testing scope

Note. This section should provide any additional information not previously covered in the System Integration Test report.

Approvals. This section should include the approvals specified in the Project Plan.

Product Packaging Information

The following example of a product description based upon the recommendations of IEEE Std 1465-1998, IEEE Standard Adoption of ISO/IEC 12119: 1994(E)—Information Technology—Software Packages—Quality Requirements and Testing [32]. Table C-24 provides an example of a product description and describes the types of information that should be present in every product description.

VERIFICATION

Inspection Log Defect Summary

The inspection log defect summary in Figure C-24 is provided as an example. This may used to record a summary of all defects identified during the inspection process. Once the inspection team has agreed to a disposition on the inspection, the leader should complete this form. An electronic copy of this log, *Inspection Log Defect Summary.doc,* is provided on the CD-ROM accompanying this book.

Inspection Log Defect Summary Description

Log Number. The recorder should enter the log number assigned to this inspection in this block (refer to Figure C-24).

Table C-24. Product description

Product description sheet SecurTracker Version 1.0

SecurTracker—An Automated Security Tool

What is SecurTracker?

SecurTracker is a *comprehensive* security management tool that incorporates the efforts of the security professionals, administrative staff, IS team and management team to integrate a complete security solution.

Architectural Goals

SecurTracker will support the five architectural goals identified below:

- Meet user requirements
- Promote system longevity
- Support incremental system implementation
- Aid iterative system refinement
- Provide mechanisms for system support and maintenance

Why Use SecurTracker?

SecurTracker is the only application available that supports collateral, SCI, and SAR environments.

SecurTracker is a multiuser application that maintains all security management data in one centralized location. Centralization offers numerous significant advantages to include the elimination of duplicate records storage, reduction of risk related to inconsistent data and a decrease in the amount of storage space (hardware costs) required to maintain security records.

SecurTracker was written using security-professional expertise. Software development teams who read security-operating manuals to determine software requirements write other competitive security management tools. This approach inevitably omits the dynamic, real-world situations that cannot be captured in an operating manual.

System Requirements

Database Server*

1 GB RAM; 60 GB Hard Drive; 1 GHz processer; Network Card, Windows NT 4.0 (SP6a)/Windows 2000 Server (SP3); Microsoft Excel; Oracle 8.1.6 or Oracle 9i.

Web Server*

256 MB RAM; 20 GB Hard Drive; 700 MHz processor; Network Card; Windows NT 4.0 (SP6a)/Windows 2000 Server (SP3 and security patches); Oracle Forms and Reports 6i; Outlook express Version 5.0/Microsoft Outlook 4.0 (with latest SP).

Web Client Machines

A Web Client machine can be any Internet-connected system that utilizes a browser; Video card supporting 256 colors, 800 x 600 resolution (or greater); Network Card.

*Although it is highly recommended that the Web and Database servers run on separate machines, a Web/Database server can operate on one machine. Using this option, the hardware requirements listed for the database server must be in place. Additionally, *all* software listed for the Database *and* Web Server specifications must be installed on this server.

Table C-24. *Continued*

Standalone Configuration

512 MB RAM; 20 GB Hard Drive; 700 MHz processor; Windows NT 4.0 (SP6a/Windows 2000 (SP3); Personal Oracle 8.1.6 or 9i; Outlook Express 5.0/Microsoft Outlook 4.0; IE 5.5.
SecurTracker is an interactive web-based solution. Client setup and maintenance is not required. Setup, configurations, and updates are accomplished at the server level, thus eliminating costly and time-consuming system maintenance.

SecurTracker Implementation Approach

SecurTracker is available as a single user stand-alone system or as a multiuser Web-based product. The user can accomplish the stand-alone installation. The installation process for multiuser implementation offers a two-phase approach; the first phase would consist of the database and Web server setup; the second phase would include data migration, validation, and testing.

The implementation team will begin by extracting the security data from the legacy system(s). Once this is accomplished, the team will then extract and the data incorporate into SecurTracker. Data extraction can be accomplished remotely, thus eliminating travel cost.

After data extraction is accomplished, the SecurTracker system will be tested to verify the accuracy of the data conversion. Any necessary adjustment to the data import functions will be made at this time. Once data has been successfully extracted from one legacy system, the extraction process can be automated for other like systems, creating an efficient repeatable process.

During the implementation phase, the implementation team will work closely with the onsite team lead and functional experts to determine an approach that ensures data integrity during migration. Final deployment includes user setup and maintenance item customization.

Data conversion, user training and phone support are available on all solutions!

Points of Contact

Susan K. Land
SecurTracker Project Manager
(256) xxx-xxxx
e-mail address

Frequently Asked Questions

Q: What languages/technologies were used in development?
A: SecurTracker is currently written in Oracle forms and reports 6i using an Oracle 9i database.

Q: Can the database be migrated to other versions of Oracle?
A: Yes. SecurTracker supports Oracle Standard and Enterprise editions of Oracle 8.1.6 or 9i.

Q: How is reporting implemented?
A: Reporting to Web, printer, e-mail are implemented by Oracle reports and leverages .PDF format.

Q: What is the hard drive footprint for the Web server?
A: 400 MB, which includes the application code.

Q: What is the hard drive footprint for the database server?
A: This depends upon the amount of space allocated for data. Initial setup is estimated at 500 MB.

Q: How is the standalone version different from the web version?
A: The standalone version will use Personal Oracle, which supports a single user. The web version uses Oracle 8i or 9i that will support any number of licensed users.

Q: Is there a limit on the number of records the system can handle?
A: The only limits occur by the size of the hard drive; therefore, by adding more physical space a needed, the number of records is limitless.

Log Number: _____   Date: _____

Software Item: _____

Moderator: _____   Review Type: _____

Defects found:

Insp.	Major Defects							Minor Defects						
	M	S	W	E	A	I	TOTAL	M	S	W	E	A	I	TOTAL

Legend:
```
     Major = a defect that would cause a problem in program operation
     Minor = all other defects
     M(issing) = required item is missing from software element
     S(tandards Compliance) = non-conformance to existing standards
     W(rong) = an error in a software element
     E(xtra) = unneeded item is included in a software element
     A(mbiguous) = an item of a software element is ambiguous
     I(nconsistent) = an item of a software element is inconsistent
```

Figure C-24. Inspection log defect summary.

Date. The recorder should enter the date of the inspection in this block.

Software Item. The recorder should enter the name of the software item that is being inspected in this book.

Moderator. The name of the inspection moderator should be entered here.

Inspection Type. The recorder should enter what type of inspection is being conducted in this block.

Defects Found. The recorder should record counts of all defects found during review of the inspection package. The defect counts should be recorded in the table as follows:

- The Insp. column will contain the name of inspector.
- The Major Defects columns (M, S, W, E, A, and I) will contain identifiers that correspond to the number of major defects found for a specific type in one of the six

possible classes. As an example: if the "W" column held the number 3, then it would mean that there were 3 major defects found that were classed as wrong.

- The first Total column will contain the total of all the major defects found for a type.
- The Minor Defects columns (M, S, W, E, A, and I) will contain numbers that correspond to the number of minor defects found for a specific type in one of the five possible classes. As an example: if the "A" column held the number 3 then it would mean that there were 3 minor defects found that were classed as Ambiguous.
- The second Total column will contain a total of all the minor defects found for a type.

Inspection Report

The inspection report form in Figure C-25 is provided as an example. This may used by the inspection leader to officially close the inspection. Once the inspection team has

Log Number: _____   Date: _____

Software Item: _____

Inspection type: _____   Duration: _____

Number of inspectors: _____   Total prep time: _____

Size of materials inspected: _____

_____

_____

Disposition: Accept _____   Conditional _____   Reinspect _____

Estimated rework effort: _____ (Hours)

Rework to be completed by : _____

Reinspection scheduled for: _____

Inspectors:

_____   _____

_____   _____

Author(s):

_____   _____

Recorder: _____

Leader: _____

Leader certification: _____   Date: _____

Additional comments: _____

Figure C-25. Inspection report form.

agreed to a disposition on an inspection, the inspection leader should complete this form. An electronic version of this report, *Inspection Report.doc,* is provided on the CD-ROM accompanying this book.

Inspection Report—Description

Log Number. The log or tracking number assigned to this inspection

Date. The date of the inspection

Software Item. The name of the software item being inspected

Inspection Type. The type of inspection being conducted

Duration. The duration of the inspection

Number of Inspectors. The number of inspectors present during the inspection

Total Prep Time. The total preparation time spent in preparation for the inspection

Size of Materials. The size of the materials inspected

Disposition. The disposition of the inspection

Estimated Rework Effort. The estimated rework effort time, in hours

Rework to Be Completed By. The estimated date of rework completion. If no rework is needed, enter "N/A."

Reinspection Scheduled For. The date for a reinspection of the software item to ensure rework completeness. It is the moderator's responsibility to decide if a reinspection is needed. If it is not needed, enter "N/A."

Inspectors. The names of the inspectors present at the inspection

Author(s). The name(s) of the author(s) responsible for the software item

Recorder. The name of the recorder present at the inspection

Leader. The leader's name

Leader Certification. The leader should sign and date this block to signal that the inspection is concluded

Additional Comments. The leader should enter additional comments as necessary for clarification.

Requirements Walk-through Form

Figure C-26 is provided as an example walk-through checklist to be used in support of the peer review of software or system requirements. This type of form may be used to resolve document issues and resolutions when performing project walk-throughs.

Software Project Plan Walk-through Checklist

Figure C-27 is provided as an example walk-through checklist to be used in support of the peer review of a software project plan. This type of form may be used to document issues and resolutions when performing project walk-throughs.

Preliminary Design Walk-through Checklist

Figure C-28 is provided as an example walk-through checklist used to support of the peer review of preliminary design. This type of form may be used to document issues and resolutions when performing project walk-throughs.

```
┌─────────────────────────────────────────────────────────────────┐
│ Software Project:                                                 │
│                                                                   │
│ Software Item(s): _____  _____  _____     │
│                                                                   │
│ Review Date: _____                                      │
│                                                                   │
│ Reviewer(s):                                                      │
│ _____  _____  _____  _____             │
│ _____  _____  _____  _____             │
│ _____  _____  _____  _____             │
│ _____  _____  _____  _____        │
│                                                                   │
│ Moderator should respond to the following:                        │
│                                              Yes    No            │
│ 1) Is their a clear understanding of the problem that this system │
│    is designed to solve?                                          │
│ 2) Are the external and internal interfaces properly defined?  ___ ___ │
│ 3) Are all the requirements traceable to the system level?     ___ ___ │
│ 4) Is prototyping conducted for customer?                      ___ ___ │
│ 5) Is the performance achievable with constraints imposed by      │
│    other system elements?                                      ___ ___ │
│ 6) Are schedule, resources, and budget consistent with            │
│    requirements?                                               ___ ___ │
│ 7) Are validation criteria complete?                           ___ ___ │
│                                                                   │
│ *If there was a "No" answer to any of the above please attach problems found and │
│ suggestions for improvement.                                      │
│                                                                   │
│ Please attach additional comments and copy of requirement items.  │
└─────────────────────────────────────────────────────────────────┘
```

Figure C-26. Requirements walk-through form.

Detailed Design Walk-through Checklist

Figure C-29 is provided as an example walk-through checklist to be used in support of the peer review of detailed design. This type of form may be used to document issues and resolutions when performing project walk-throughs.

Program Code Walk-through Checklist

Figure C-30 is provided as an example walk-through checklist to be used in support of the peer review of developed code. This type of form may be used to document issues and resolutions when performing project walk-throughs. *Note:* the checklist assumes that a design walk-through has already been done.

Test Plan Walk-through Checklist

Figure C-31 is provided as an example walk-through checklist to be used in support of the peer review of a software test plan. This type of form may be used to document issues and resolutions when performing project walk-throughs.

Software Project:

Software Item(s): _____ _____ _____

Review Date: _____

Reviewer(s):

_____ _____ _____ _____
_____ _____ _____ _____
_____ _____ _____ _____
_____ _____ _____ _____

Moderator please respond to the following:

                                                                    Yes   No
1) Is the software scope unambiguously defined and bounded?   ____   ____
2) Is the terminology clear?   ____   ____
3) Are the resources adequate for scope?   ____   ____
4) Are the resources readily available?   ____   ____
5) Are the tasks properly defined and sequenced?   ____   ____
6) Is task parallelism reasonable given the available resources?   ____   ____
7) Have both historical productivity and quality data been used?   ____   ____
8) Have any differences in estimates been reconciled?   ____   ____
9) Are the preestablished budgets and deadlines realistic?   ____   ____
10) Is the schedule consistent?   ____   ____

* If there was a "No" answer to any of the above, please attach problems found and suggestions for improvement.

Please attach additional comments and a copy of requirement items.

Figure C-27. Software project plan walk-through checklist.

Walk-through Summary Report

Figure C-32 is provided as an example walk-through summary report. This may be used to document issues and resolutions when performing peer reviews.

Classic Anomaly Class Categories

Anomaly classes provide evidence of nonconformance and may be categorized. The following examples are provided:

Missing
Extra (superfluous)
Ambiguous
Inconsistent
Improvement desirable

Software Project: _____

Software Item(s): _____

Review Date: _____

Reviewer(s):

_____   _____
_____   _____
_____   _____

Moderator please respond to the following:

	Yes	No
1) Are software requirements reflected in the software architecture?	___	___
2) Is effective modularity achieved? Are modules functionally independent?		
3) Is the program architecture factored?	___	___
4) Are interfaces defined for modules and external system elements?		
5) Is the data structure consistent with information domain?	___	___
6) Is the data structure consistent with software requirements?	___	___
7) Has maintainability been considered?	___	___

*If there was a "No" answer to any of the above, please attach problems found and suggestions for improvement.

Please attach additional comments and a copy of requirement items.

Figure C-28. Preliminary design walk-through checklist.

Not conforming to standards

Risk-prone, that is, the review finds that although an item was not shown to be "wrong," the approach taken involves risks (and there are known safer alternative methods)

Factually incorrect

Not implementable (e.g., because of system constraints or time constraints)

Editorial

VALIDATION

Example Test Classes

Check for Correct Handling of Erroneous Inputs

Test Objective. Check for proper handling of erroneous inputs: characters that are not valid for this field, too many characters, not enough characters, value too large, value too small, all selections for a selection list, no selections, or all mouse but-

Software Project:

Software Item(s): _____

Review Date: _____

Reviewer(s):

_____   _____
_____   _____
_____   _____
_____   _____

Moderator please respond to the following:

	Yes	No
1) Does the algorithm accomplish the desired function?	___	___
2) Is the algorithm logically correct?	___	___
3) Is the interface consistent with the architectural design?	___	___
4) Is logical complexity reasonable?	___	___
5) Has error handling been specified and built-in?	___	___
6) Is local data structure properly defined?	___	___
7) Are structured programming constructs used throughout?	___	___
8) Is design detail amenable to the implementation language?	___	___
9) Which are used: operating system or language-dependent features?		
10) Has maintainability been considered?	___	___

*If there was a "No" answer to any of the above, please attach problems found and suggestions for improvement.

Please attach additional comments and a copy of requirement items.

Figure C-29. Detailed design walk-through checklist.

tons clicked or double clicked all over the client area of the item with focus. Test class # xx.

Validation Methods Used: Test.

Recorded Data. User action or data entered, screen/view/dialog/control with focus, resulting action.

Data Analysis. Was resulting action within general fault handling defined capabilities in the [PROJECT ABBREVIATION] SysRS and design in [PROJECT ABBREVIATION] SDD?

Assumptions and Constraints. None.

Check for Maximum Capacity

Test Objective. Check software and database maximum capacities for data: enter data until maximum number of records specified in the design is reached for each table; operate program and add one more record. Test class # xx.

Software Project:

Software Item(s): _____

Review Date: _____

Reviewer(s):

_____   _____

_____   _____

_____   _____

Moderator please respond to the following:

	Yes	No
1) Is design properly translated into code?	____	____
2) Are there misspellings or typos?	____	____
3) Has proper use of language conventions been made?	____	____
4) Is there compliance with coding standards for language style, comments, module prologue?	____	____
5) Are incorrect or ambiguous comments present?	____	____
6) Are typing and data declaration proper?	____	____
7) Are physical contents correct?	____	____
8) Have all items on the design walk-through checklist been reapplied (as required)?	____	____

*If there was a "No" answer to any of the above, please attach problems found and suggestions to improvement.

Please attach additional comments and a copy of requirement items.

Figure C-30. Program code walk-through checklist.

Validation Methods Used: Test.

Recorded Data. Record number of records in each table, resulting actions.

Data Analysis. Was resulting action to the maximum plus one normal?

Assumptions and Constraints. This test requires someone to create through some method a populated database with records several times more than what actually exists in the sample data set. This takes a good deal of time. Integration and qualification test only.

User Interaction Behavior Consistency

Test Objective. Is the interaction behavior of the user interface consistent across the application or module under test? Tab through controls, using mouse click and double click on all controls and in null area, maximize, minimize, normalize, switch focus to another application and then back, update data on one view that is included on another and check to see if the other view is updated when data is saved on the first view. Use function keys and movement keys and other standard key combina-

---

Software Project:

Software Item(s): _____
Review Date: _____

Reviewer(s):

_____   _____
_____   _____
_____   _____
_____   _____

Moderator please respond to the following:

                                                             Yes   No

1) Have major test activities been properly identified and
    sequenced?   _____ _____
2) Has traceability to validation criteria/requirements been
    established as part of software requirements analysis?   _____ _____
3) Are major functions demonstrated early?
4) Is the test plan consistent with the overall project plan?   _____ _____
5) Has a test schedule been explicitly defined?   _____ _____
6) Are test resources and tools identified and available?   _____ _____
7) Has a test record keeping mechanism been established?   _____ _____
8) Have test drivers and stubs been identified, and has work
    to develop them been scheduled?   _____ _____
9) Has stress testing for software been specified?   _____ _____

*If there was a "No" answer to any of the above, please attach problems found and
suggestions for improvement.

Please attach additional comments and a copy of requirement items.

Figure C-31. Test plan walk-through checklist.

tions (clipboard combos, control key windows and program defined sets, Alt key defined sets). Enter invalid control and Alt key sets to check for proper handling. Test class # xx.

Validation Methods Used: Test, Inspection.

Recorded Data. Record any anomalies of action resulting from user action not conforming to the behavioral standards for windows programs.

Data Analysis. Was resulting action within behavioral standards of windows programs as defined in the [PROJECT ABBREVIATION] SysRS and design in [PROJECT ABBREVIATION] SDD? Was behavior consistent across the application or module as defined in the [PROJECT ABBREVIATION] SysRS and design in [PROJECT ABBREVIATION] SDD?

Assumptions and Constraints. If testing at module level, the multiple view portion of the test may not apply due to having only a single view.

Project Name:	Date:
Element Reviewed:	
Review Team:	
Problems Found:	
Proposed Solutions:	
Comments:	
Action Taken:	

Figure C-32. Walk-through summary report.

Retrieving Data

Test Objective. Is the data retrieved correct: for each dialog, list box, combo box, and other controls that show lists? Check the data displayed for correctness. Test class # xx.

Validation Methods Used: Test, Inspection.

Recorded Data. Record data displayed and data sources (records from tables, resource strings, code sections).

Data Analysis. Was data displayed correctly? Compare data displayed with sources.

Assumptions and Constraints. Requires alternate commercial database software to get records from the database.

Saving Data

Test Objective. Is the data entered saved to the database correctly? For each dialog, list box, combo box and other controls that show lists check the data entered and saved for correctness in the database. Test class # xx.

Validation Methods Used: Test, Inspection.

Recorded Data. Record data entered and data destinations (records from tables).

Data Analysis. Was data saved correctly? Compare data entered with destination.

Assumptions and Constraints. Requires alternate commercial database software to get records from the database.

Display Screen and Printing Format Consistency

Test Objective. Are user interface screens organized and labeled consistently? Are printouts formatted as specified? Enter data to maximum length of field in a printout and then print. Show all screens (views, dialogs, print previews, OLE devices) and dump their image to paper. Test class # xx.

Validation Methods Used: Inspection.

Recorded Data. Screen dumps and printouts.

Data Analysis. Was the printout format correct? Were the fields with max length data not clipped? Were the labels and organization of screens consistent across the application or module as defined in the [PROJECT ABBREVIATION] SDD?

Assumptions and Constraints. The module that performs forms printing is required with all other modules during their testing.

Check Interactions Between Modules

Test Objective. Check the interactions between modules. Enter data and save it in one module and switch to another module that uses that data to check for latest data entered. Switch back and forth between all of the modules while manipulating data and check for adverse results or program faults. Test class # xx.

Validation Methods Used: Demonstration.

Recorded Data. Screen dumps.

Data Analysis. Was resulting actions within specifications as defined in the [PROJECT ABBREVIATION] SysRS and design in [PROJECT ABBREVIATION] SDD?

Assumptions and Constraints. Requires customer participation. Requires all modules and supporting software.

Measure Time of Reaction to User Input

Test Objective. Check average response time to user input action: clock time from (saves, retrieves, dialogs open and closes, views open and closes), clock time from any response to user action that takes longer than 2 seconds. Test class # xx.

Validation Methods Used: Test, Analysis.

Recorded Data. Record action and response clock time. Organize into categories and average their values. Are all average values less than the minimum response time specified?

Data Analysis. Organize into categories and average their values. Are all average values less than the minimum response time specified as defined in the [PROJECT ABBREVIATION] SysRS and design in [PROJECT ABBREVIATION] SDD?

Assumptions and Constraints. None.

Functional Flow

Test Objective. Exercise all menu, buttons, hotspots, and so on that cause a new display (view, dialog, OLE link) to occur. Test class # xx.

Validation Methods Used: Demonstration.

Recorded Data. Screen Dumps.

Data Analysis. Were resulting actions with specifications as defined in the [PROJECT ABBREVIATION] SysRS and design in [PROJECT ABBREVIATION] SDD?

Assumptions and Constraints. Requires customer participation. Requires all modules and supporting software.

Examples of System Testing

Functional Testing. Testing against operational requirements is Functional Testing. This section should provide references to the SysRS to show a complete list of functional requirements and resulting test strategy.

Performance Testing. Testing against performance requirements is Performance Testing. This section should provide references to the SysRS to show a complete list of performance requirements and resulting test strategy.

Reliability Testing. Testing against reliability requirements is Reliability Testing. This section should provide references to the SysRS to show a complete list of reliability requirements and resulting test strategy.

Configuration Testing. Testing under different hardware and software configurations to determine an optimal system configuration is Configuration Testing. This section should provide references to the SysRS to show a complete list of configuration requirements and resulting test strategy.

Availability Testing. Testing against its operational availability requirements is Availability Testing. This section should provide references to the SysRS to show a complete list of availability requirements and resulting test strategy.

Portability Testing. Testing against portability requirements is Portability Testing. This section should provide references to the SysRS to show a complete list of portability requirements and resulting test strategy.

Security and Safety Testing. Testing against security and safety requirements is Security and Safety Testing. This section should provide references to the SysRS to show a complete list of security and safety requirements and resulting test strategy.

System Usability Testing. Testing against usability requirements is System Usability Testing. This section should provide references to the SysRS to show a complete list of system usability requirements and resulting test strategy.

Internationalization Testing. Testing against internationalization requirements is Internationalization Testing. This section should provide references to the SysRS to show a complete list of internationalization requirements and resulting test strategy.

Operations Manual Testing. Testing against the procedures in the operations manual to see if it can be operated by the system operator is Operations Manual Testing. This section should provide references to the operations manual to show a complete list of operator requirements and resulting test strategy.

Load Testing. Testing that attempts to cause failures involving how the performance of a system varies under normal conditions of utilization is Load Testing. This section should provide the test strategy.

Stress Testing. Testing that attempts to cause failures involving how the system behaves under extreme but valid conditions (e.g., extreme utilization, insufficient memory, inadequate hardware, and dependency on overutilized shared resources) is Stress Testing. This section should provide the test strategy.

Robustness Testing. Testing that attempts to cause failures involving how the system behaves under invalid conditions (e.g., unavailability of dependent applications, hardware failure, and invalid input such as entry of more than the maximum amount of data in a field) is Robustness Testing. This section should provide the test strategy.

Contention Testing. Testing that attempts to cause failures involving concurrency is Contention Testing. This section should provide the test strategy.

Test Design Specification

This example test design specification is based upon IEEE-Std 829, IEEE Standard for Software Test Documentation. [5]

Purpose. This section should provide an overview description the test approach and identify the features to be tested by the design.

Outline. A test design specification should have the following structure:

1. *Test design specification identifier.* Specify the unique identifier assigned to this test design specification as well as a reference to any associated test plan.
2. *Features to be tested.* Identify all test items describing the features that are the target of this design specification. Supply a reference for each feature to any specified requirements or design.

3. *Approach refinements.* Specify any refinements to the approach and any specific test techniques to be used. The method of analyzing test results should be identified (e.g., visual inspection). Describe the results of any analysis that provides a rationale for test case selection. For example, a test case used to determine how well erroneous user inputs are handled.

4. *Test identification.* Include the unique identifier and a brief description of each test case associated with this design. A particular test case may be associated with more than one test design specification.

5. *Feature pass/fail criteria.* Specify the criteria to be used to determine whether the feature has passed or failed.

The sections should be ordered in the specified sequence. Additional sections may be included at the end. If some or all of the content of a section is in another document, then a reference to that material may be listed in place of the corresponding content. The referenced material must be attached to the test design specification or readily available to users.

Test Case Specification

This example test case specification is based upon IEEE-Std 829, IEEE Standard for Software Test Documentation. [5]

Purpose. To define a test case identified by a test design specification.

Outline. A test case specification should have the following structure:

1. *Test case specification unique identifier.* Specify the unique identifier assigned to this test case specification.

5. *Test items.* Identify and briefly describe the items and features to be exercised by this test case. For each item, consider supplying references to the following test item documentation: Requirements specification, design specification, users guide, operations guide, and installation guide.

6. *Input specifications.* Specify each input required to execute the test case. Some of the inputs will be specified by value (with tolerances where appropriate), whereas others, such as constant tables or transaction files, will be specified by name. Specify all required relationships between inputs (e.g., timing).

7. *Output specifications.* Specify all of the outputs and features (e.g., response time) required of the test items. Provide the exact value (with tolerances where appropriate) for each required output or feature.

8. *Environmental needs.* Specify all hardware and software requirements associated with this test case. Include facility or personnel requirements if appropriate.

9. *Special procedural requirements.* Describe any special constraints on the test procedures that execute this test case.

10. *Intercase dependencies.* List the identifiers of test cases that must be executed prior to this test case. Summarize the nature of any dependencies.

If some or all of the content of a section is in another document, then a reference to that material may be listed in place of the corresponding content. The referenced material must be attached to the test case specification or readily available to users of the case specification.

A test case may be referenced by several test design specifications, and enough specific information must be included in the test case to permit reuse.

Test Procedure Specification

This example test procedure specification is based upon IEEE-Std 829, IEEE Standard for Software Test Documentation. [5]

Purpose. To document the steps in support of test set execution or, more generally, the steps used to evaluate a set of features associated with a software item.

Outline. A test procedure specification should have the following structure:

1. *Test procedure specification identifier.* The unique identifier assigned to the test procedure specification. A reference to the associated test design specification should also be provided.
2. *Purpose.* Describe the purpose of this procedure providing a reference for each executed test case. In addition, provide references to relevant sections of the test item documentation (e.g., references to usage procedures) where appropriate.
3. *Special requirements.* Identify any special requirements that are necessary for the execution. These may include prerequisite procedures, required skills, and environmental requirements.
4. *Procedure steps.* Describe the procedural steps supporting the test in detail, the following provides guidance regarding the level of detail required:

    Log. Describe the method or format for logging the results of test execution, the incidents observed, and any other test events.

    Setup. Describe the sequence of actions necessary to prepare for execution of the procedure.

    Start. Describe the actions necessary to begin execution of the procedure.

    Proceed. Describe any actions necessary during execution of the procedure.

    Measure. Describe how the test measurements will be made.

    Shutdown. Describe the actions necessary to immediately suspend testing if required.

    Restart. Identify any restart points and describe the actions necessary to restart the procedure at each of these points.

    Stop. Describe the actions necessary to bring execution to an orderly halt.

    Wrap up. Describe the actions necessary to restore the environment.

    Contingencies. Describe the actions necessary to deal with anomalous events that may occur during execution.

The sections should be ordered as described. Additional sections, if required, may be included at the end of the specification. If some or all of the content of a section is in another document, then a reference to that material may be listed in place of the corresponding content. The referenced material must be attached to the test procedure or readily available to users of the procedure specification.

Test Item Transmittal Report

This example test item transmittal report is based upon IEEE-Std 829, IEEE Standard for Software Test Documentation [5].

Purpose. To identify all items being transmitted for testing. This report should identify the person responsible for each item, the physical location of the item, and item status. Any variations from the current item requirements and designs should be noted in this report.

Outline. A test item transmittal report should have the following structure:

1. *Transmittal reports identifier.* The unique identifier assigned to this test item transmittal report.
2. *Transmitted items.* Describe all test items being transmitted, including their version/revision level. Supply references associated with the documentation of the item including, but not necessarily limited to, the test plan relating to the transmitted items. Indicate the individuals responsible for the transmitted items.
3. *Location.* Identify the location of all transmitted items. Identify the media that contain the items being transmitted. Indicate media identification and labeling.
4. *Status.* Describe the status of the test items being transmitted, listing all incident reports relative to these transmitted items.
5. *Approvals.* List all transmittal approval authorities, providing a space for associated signature and date.

The sections should be ordered as described. Additional sections, if required, may be included at the end of the specification. If some or all of the content of a section is in another document, then a reference to that material may be listed in place of the corresponding content. The referenced material must be attached to the test item transmittal report or readily available to users of the transmittal report.

Test Log

This example test log is based upon IEEE-Std 829, IEEE Standard for Software Test Documentation [5].

Purpose. To document the chronological record of test execution.

Outline. A test log should have the following structure:

1. *Test log identifier.* The unique identifier assigned to this test log.
2. *Description.* This section should include the identification of the items being tested, environmental conditions, and type of hardware being used.
3. *Activity and event entries.* Describe event start and end activities recording the occurrence date, time, and author. A description of the event execution, all results, as-

sociated environmental conditions, anomalous events, and test incident report number should all be included as supporting information.

The sections should be ordered in the specified sequence. Additional information may be included at the end. If some or all of the content of a section is in another document, then a reference to that material may be listed in place of the corresponding content. The referenced material must be attached to the test log or readily available to users of the log.

Test Incident Report

This example test incident report is based upon IEEE-Std 829, IEEE Standard for Software Test Documentation [5].

Purpose. To document all events occurring during the testing process requiring further investigation.

Outline. A test incident report should have the following structure:

1. *Test incident report identifier.* The unique identifier assigned to this test incident report.
2. *Summary.* Provide a summary description of the incident identifying all test items involved and their version/revision level(s). References to the appropriate test procedure specification, test case specification, and test log should also be supplied.
3. *Incident description.* Provide a description of the incident. This description should include the following items: inputs and anticipated results, actual results, all anomalies, the date and time and procedure step, a description of the environment, any attempt to repeat, and the names of testers and observers.
4. *Related activities and observations* that may help to isolate and correct the cause of the incident should be included (e.g., describe any test case executions that might have a bearing on this particular incident and any variations from the published test procedure).
5. *Impact.* Describe the impact this incident will have on existing test plans, test design specifications, test procedure specifications, or test case specifications.

The sections should be ordered in the sequence described above. Additional sections may be included at the end if desired. If some or all of the content of a section is in another document, then a reference to that material may be listed in place of the corresponding content. The referenced material must be attached to the test incident report or made readily available to users of the incident report.

Test Summary Report

This example test summary report is based upon IEEE-Std 829, IEEE Standard for Software Test Documentation [5].

Purpose. To provide a documented summary of the results of testing activities and associated result evaluations.

Outline. A test summary report should have the following structure:

1. *Test summary report identifier.* Include the unique identifier assigned to this test summary report.
2. *Summary.* Provide a summary of items tested. This should include the identification of all items tested, their version/revision level, and environment in which the testing activities took place. For each test item, also supply references to the following documents if they exist: test plan, test design specifications, test procedure specifications, test item transmittal reports, test logs, and test incident reports.
3. *Variances.* Report any variances of the test items from their specifications associating the reason for each variance.
4. *Comprehensiveness assessment.* Identify any features not sufficiently tested, as described in the test plan, and explain the reasons.
5. *Summary of results.* Provide a summary of test results including all resolved and unresolved incidents.
6. *Evaluation.* This overall evaluation should be based upon the test results and the item level pass/fail criteria. An estimate of failure risk may also be included.
7. *Summary of activities.* Provide a summary of all major testing activities and events.
8. *Approvals.* List the names and titles of all individuals who have approval authority. Provide space for the signatures and dates.

The sections should be ordered in the specified sequence. Additional sections may be included just prior to approvals if required. If some or all of the content of a section is in another document, then a reference to that material may be listed in place of the corresponding content. The referenced material must be attached to the test summary report or readily available to users.

ORGANIZATIONAL PROCESS FOCUS

Organizational Process Improvement Checklist

The IDEAL (initiating, diagnosing, establishing, acting, and learning) model is an organizational improvement model that was originally developed to support CMM®-based software process improvement [56]. It serves as a roadmap for initiating, planning, and implementing improvement actions. This model can serve to lay the groundwork for a successful improvement effort (initiate), determine where an organization is in reference to where it may want to be (diagnose), plan the specifics of how to reach goals (establish), define a work plan (act), and apply the lessons learned from past experience to improve future efforts (learn).

This model serves as the basis for CMMI® process improvement. The information presented in Table C-25 provides a matrix of IDEAL[SM] phases, activities, details, and deliverables in support of process improvement and may be used as a reference checklist in support of process improvement activities.

Table C-25. Organizational process improvement checklist using IDEAL[SM] [65]

IDEAL[SM] Phases	IDEAL[SM] Activities	IDEAL[SM] Details	Organization milestone details and deliverables
Initiating	Stimulus for Change	Know what prompted this particular change: • Reaction to unanticipated events/circumstances • Edict from "on high" • Enactment of change as part of proactive continuous improvement approach	Business results, problem reports
	Set Context	Identify where this change fits within the larger organizational context: • Within the organization's business strategy Specify business goals and objectives that will be realized or supported: • The organization's core mission • Business goals and objectives • A coherent vision for the future • A strategy for achieving the vision • Reference models Identify impact on other initiatives and on-going work	Business goals and objectives
	Build Sponsorship	Secure the support required to give the change a reasonable chance of succeeding Effective sponsors: • Give personal attention to the effort • Stick with the change through difficult times • Commit scarce resources to the effort • Change their own behavior to be consistent with the change being implemented • Modify the reward system.	Sponsor(s) signed up
	Charter Infrastructure	Adjust relevant organizational systems to support the effort Establish an infrastructure for managing specifics of implementation • An oversight group (senior-level people) • A change agency group (staff to the oversight body) • One or more technical working groups (TWGs)	Engineering Process Group (EPG) Charter
Diagnosing	Characterize Current and Desired States	Identify the organization's current state with respect to a related standard or reference model: • A reference model (e.g., capability maturity model) • Standardized appraisal instruments and methodologies associated with the model • Business objectives aligned with the change effort	Process Appraisal Plan, Assessment Plan

Table C-25. *Continued*

IDEAL^{SM} Phases	IDEAL^{SM} Activities	IDEAL^{SM} Details	Organization milestone details and deliverables
Diagnosing (*cont.*)	Characterize Current and Desired States (*cont.*)	Identify the organization's desired state against the same standard or model: • Analyze the gap between the current and desired states • Determine aspects of current state to retain • Determine the work to reach the desired state • Align recommendations with business objectives • Identify potential projects where you may pilot or test	
	Develop Recommend-ations	Recommend actions that will move the organization from where it is to where you want it to be Identify potential barriers to the effort	Recommendation report
Establishing	Set Priorities	Develop priorities among the recommended actions on the basis of such things as: • Availability of key resources • Dependencies between the actions recommended • Funding Organizational realities: • Resources needed for change are limited • Dependencies exist between recommended activities • External factors may intervene • Organization's strategic priorities must be honored	
	Develop Approach	Develop an approach that reflects priorities and current realities: • Strategy to achieve vision (set during initiating phase) • Specifics of installing the new technology • New skills and knowledge required of the people who will be using the technology • Organizational culture • Potential pilot projects • Sponsorship levels • Market forces	
	Plan Actions	Develop plans to implement the chosen approach: • Deliverables, activities, and resources • Decision points • Risks and mitigation strategies • Milestones and schedule • Measures, tracking, and oversight	Organizational Process Improvement Plan, Process Action Plan (PAP) *(continued)*

Table C-25. *Continued*

IDEAL^{SM} Phases	IDEAL^{SM} Activities	IDEAL^{SM} Details	Organization milestone details and deliverables
Acting	Create Solution	Bring together everything available to create a "best guess" solution specific to organizational needs, for example: • Existing tools, processes, knowledge, skills • New knowledge, information • Outside help Solution should: • Identify performance objectives • Finalize the pilot/test group • Construct the solution material and training with pilot/test group • Develop draft plans for pilot/test and implementation Technical working groups develop the solution with pilot/test group	Revised process description and/or process asset
	Pilot/Test Solution	Put the solution in place on a trial basis to learn what works and what doesn't: • Train the pilot or test group • Perform the pilot or test • Gather feedback	Process lessons learned
	Refine Solution	Revisit and modify the solution to incorporate new knowledge and understanding Iterations of pilot/test and refine may be necessary before arriving at a solution that is deemed satisfactory	Revised process description and/or process asset
	Implement Solution	When solution is deemed workable and sufficient, implement it throughout the organization: • Adjust the plan as necessary • Execute the plan • Conduct post-implementation analysis • Various approaches may be used: top-down, lateral, or staged	Process Lessons Learned
Learning	Analyze and Validate	Determine what was actually accomplished by the effort: • In what ways did it/did it not accomplish its intended purpose? • What worked well? • What could be done more effectively or efficiently? Compare what was accomplished with the intended purpose for undertaking the change Summarize/roll up lessons learned regarding processes used to implement IDEAL	Change in measurements

Table C-25. *Continued*

IDEAL^{SM} Phases	IDEAL^{SM} Activities	IDEAL^{SM} Details	Organization milestone details and deliverables
Learning (*cont.*)	Propose Future Actions	Develop recommendations concerning management of future change efforts using IDEAL: • Improving an organization's ability to use IDEAL • Address a different aspect of the organization's business	

Organization Process Appraisal Checklist

The information provided in Table C-26 may be used in support of the organization's process appraisals. An electronic version of this checklist, identified as *Process Appraisal Checklist.doc,* is provided on the CD-ROM accompanying this book.

Process Lessons Learned

The documentation of lessons learned facilitates individual project knowledge sharing so as to benefit the entire organization. A successful lessons-learned program can support the repeat of desirable outcomes and avoid undesirable outcomes. Lessons learned should not only be used as a closeout activity; the continual recording of lessons learned throughout the project is the best way to ensure that information is accurately recorded. Figure C-33 a lessons-learned template.* Please refer to the CD-ROM accompanying this book for an electronic version of this template entitled *Lessons Learned.doc.*

ORGANIZATIONAL PROCESS DEFINITION

Organizational Policy Examples

The following information is provided for illustrative purposes. Each organization should focus on the development of policies and practices that best support its needs. Each organizational policy should include a top-level statement, any relevant responsibilities, and associated specific actions as desired. The following provides three examples of an organizational policy in support of configuration management activities.

Example 1

Individuals responsible for project leadership shall ensure that all software development, system integration, and system engineering projects conduct configuration management activities to maintain the integrity of their products and data. Each project's configuration management effort shall ensure:

---

*Adapted from a template written by Covansys for the City of Raleigh, NC, Enterprise PMO. Generic format by CVR-IT (www.cvr-it.com). May be used freely if referenced.

- The identification of configuration items (CIs) and associated baselines
- The use of a change management process for managing changes to baselines
- The support of process audits
- Configuration status tracking and reporting

Example 2

Our goal is to deliver the highest quality products to our customers. To ensure that the integrity of a project's products is established and maintained throughout the project's life cycle, this configuration management policy has three primary requirements:

1. Each Project must have a CM Lead to develop and carry out the CM process. This individual is responsible for defining, implementing, and maintaining the CM Plan.

2. Every Project must have a CM Plan identifying the products to be managed by the CM process, the tools to be used, the responsibilities of the CM team, and the procedures to be followed. The CM Plan must be prepared in a timely manner.

Table C-26. Organization process appraisal checklist

Checklist	Dates + resources

1. Identify Appraisal Scope
   a. Appraisal input subject to sponsor approval
2. Develop Appraisal Plan (using this checklist)
   a. Select appraisal class
   b. Identify appraisal scope
   c. Select appraisal team members
   d. Develop and publish appraisal plan
3. Prepare Organization for Appraisal
   a. Prepare and train appraisal team
   b. Brief appraisal participants
   c. Data collection and analysis
   d. Administer questionnaire
   e. Conduct initial document review
   f. Remediate, consolidate, and triage responses
   g. Conduct opening meeting
   h. Interview project leaders
   i. Consolidate information
   j. Interview middle managers
   k. Consolidate information
   l. Interview FAR groups
   m. Consolidate information
   n. Verify and validate objective evidence
   o. Prepare draft findings
   p. Present draft findings
   q. Consolidate information
4. Rate and prep final findings
5. Present final findings
6. Conduct executive session
7. Present final findings to appraisal participants
8. Wrap-up appraisal	

[Project Name] Lessons Learned
Revision [xx], date

Team Meeting

During each project team meeting discuss what strategies contributed to success as well as areas of potential improvement. Enter your conclusions in the table below (insert rows as needed):

Strategies and Processes that led to Success

Date   Description

Areas of Potential Improvement

Date   Description

Project Closeout Discussion

All stakeholders should participate in the lessons learned closeout discussion. Use the questions below to trigger discussion and summarize findings.

1. List the project's three biggest successes:

Description   Contributing Factors

2. Other relevant successes:

Description   Contributing Factors

3. List improvement items and proposed strategies:

Description   Improvement Strategy

4. Additional Comments

Document Approval

The following stakeholders have reviewed the information contained in this project lessons-learned document and agree with its content:

Name   Title   Signature   Date

Figure C-33. Process lessons learned template.

3. Every Project must have a Configuration Control Board (CCB). The CCB authorizes the establishment of product baselines and the identification of configuration items/units. It also represents the interests of the Project Manager and all groups who may be affected by changes to the configuration. It reviews and authorizes changes to controlled baselines and authorizes the creation of products from the baselines.

Example 3

In order to ensure that delivered products contain only high-quality content, engineering projects shall:

- Establish and maintain baselines of all identified work products
- Track and control changes to all identified work products
- Establish and maintain the integrity of all baselines

Process Definition Form

A defined process should clearly state all process inputs, entry criteria, associate activities, roles, measures, required verification, outputs, and exit criteria. Figure C-35 provides an example of a form that may be used in support of process definition and development.

Process Asset Library Catalog

IEEE Std 610.12-1990 (R2002), IEEE Standard Glossary of Software Engineering Terminology, defines process as, "A set of activities. A sequence of steps performed for a given purpose; for example, the software development process" [2]. The CMMI defines process as, "... activities that can be recognized as implementations of practices in a CMMI model. These activities can be mapped to one or more practices in CMMI process areas to allow a model to be useful for process improvement and process appraisal" [61].

The CMMI further defines Organizational Process Assets as, "... artifacts that relate to describing, implementing, and improving processes (e.g., policies, measurements, process descriptions, and process implementation support tools). The term 'process assets' is used to indicate that these artifacts are developed or acquired to meet the business objectives of the organization, and they represent investments by the organization that are expected to provide current and future business value" [61].

The organizational process assets that are described in CMMI models include the following [61]:

- Organization's set of standard processes, including the process architectures and process elements
- Descriptions of life-cycle models approved for use
- Guidelines and criteria for tailoring the organization's set of standard processes
- Organization's measurement repository
- Organization's process asset library

An organization's process asset library may, therefore, be described as a library of information used to store process assets that may be used in the support of the definition,

OVERVIEW							

Entry Criteria	Exit Criteria

Inputs	Outputs

Required Activities

Stakeholders L = lead; S = support; R = review; O = optional (as necessary)							

Measures

Organizational Process Improvement Information

Verification

Tailoring

Implementation Guidance

Supporting Documentation and Assets

Figure C-34. Process definition form [50].

implementation, and management of processes within an organization. Some examples of this type of documentation include policies, processes, procedures, checklists, lessons learned, templates and plans, referenced standards, and training materials.

Measures Definition for the Organizational Processes

IEEE Std. 982.1-1998, IEEE Standard Dictionary of Measures to Produce Reliable Software; IEEE 1061-1998, IEEE Standard for a Software Quality Metrics Methodology;

IEEE Std. 14143.1-2000, Implementation Note for IEEE Adoption of ISO/IEC 14143-1:1998, Information Technology—Software Measurement—Functional Size Measurement—Part 1: Definition of Concepts; ISO/IEC 15939:2001, Information Technology—Software Engineering—Software Measurement Process; and CMMI® materials were used as to develop the following suggestions for measures in support of organizational processes.

Metrics are often classified as either process or product metrics. Process metrics are based upon characteristics associated with the development process or environment. Product metrics are measures associated with the software product being developed. Examples of these types of metrics are provided in Table C-27.

The basic information in support of process measurement that should be collected is [55]:

- Released defect levels
- Product development cycle time
- Schedule and effort estimating accuracy
- Reuse effectiveness
- Planned and actual cost

ORGANIZATIONAL TRAINING

Training Request Form

Figure C-35 is a form that may be used to request training. Project managers should note training needs within their project. Use as many forms as required to document all training needs. Training managers may use this form to analyze, evaluate, validate and prioritize training requests. An electronic version, *Training Request Form.doc,* is provided on the CD-ROM accompanying this book.

Training Log

Figure C-36 a form may be used to document training. Project managers, or individuals delegated with this responsibility, should record all completed training. Training managers may use this form to analyze, evaluate, validate, and prioritize training requests. An electronic version, *Training Log.doc,* is included on the CD-ROM accompanying this book.

Table C-27. Process versus product metrics

Process Metrics
Estimated and actual duration (calendar time): track for individual tasks, project milestones, and process support functions

Estimated and actual effort (labor hours): track for individual tasks, project milestones, and process support functions.

Product Metrics
Count lines of code, function points, object classes, numbers of requirements
Time spent in development and maintenance activities
Count the number of defects identifying their type, severity, and status

Project Name: _____ Project Manager: _____

Project Location: _____ Date: _____

Will project contract pay for requested training?   Yes   No   Labor Only

Describe the training requirement:

Purpose and date training required:

Number of employees to be trained:

Can employees be released for training during work hours?   Yes   No
Are subject matter experts available?   Yes   No
If so, Names:

Project training requirement priority (5 is the highest, 1 is the lowest)

5   4   3   2   1

Recommended provider (if known)

Recommended course title (if known)

Recommended action:

Program Manager: _____ Date: _____

Division Manager: _____ Date: _____

Status: (To be completed by Training Manager)

Training Manager: _____ Date: _____

SBU Training Requirement Priority (5 is the highest, 1 is the lowest)

5   4   3   2   1

ACTION TAKEN:

Invalid requirement* _____ Valid requirement_____

Placed on training schedule_____ No action taken*_____

Funding identified: _____

Training course identified: _____

Placed on development schedule: _____

Provider: _____

Course title: _____ Date: _____

*Remarks:

Figure C-35. Training request form.

Trainee Name	Date	Location	Description of Training	No. Hours	ID #	Comments

Figure C-36. Training log.

INTEGRATED PROJECT MANAGEMENT

Status Reviews

Reviews should occur throughout the life cycle as identified in the Software Project Management Plan. These reviews can be conducted as internal management or customer management reviews. The purpose of these types of reviews is to determine the current status and risk of the software effort. The data presented during these reviews should be used as the basis for decision making regarding the future path and progress of the project.

Internal Management Reviews. The following are suggested agenda items for discussion during an internal management review:

- An overview of current work to date
- Status versus project plan comparison (cost and schedule)
- Discussion of any suggested alternatives (optional)
- Risk review and evaluation
- Updates to the Project Management Plan (if applicable)

Customer Management Reviews. The following are suggested agenda items for discussion during a customer management review:

- An overview of current work to date
- Status versus project plan comparison (cost and schedule)
- Discussion of any suggested alternatives (optional)
- Risk review and evaluation
- Customer approval of any updates to the project plan

Critical Dependencies Tracking

IEEE Std 1490-2003, IEEE Adoption of PMI Standard, A Guide to the Project Management Body of Knowledge, describes the identification, tracking, and control of all items critical to the successful management of a project. The PMI refers to the identification and tracking of critical dependencies as integrated change control. According to the IEEE Std 1490:

> Integrated change control is concerned with (a) influencing the factors that create changes to ensure that changes are agreed upon, (b) determining that a change has occurred, and (c) managing the actual changes when and as they occur. The original defined project scope and the integrated performance baseline must be maintained by continuously managing changes to the baseline, either by rejecting new changes or by approving changes and incorporating them into a revised project baseline [34].

As described by IEEE Std 1490, there are three key inputs that support integrated change control. The content of the software project management plan provides key input and becomes the change control baseline. Status reporting provides key input of interim information on project performance. Last, change requests provide a record of the changes requested to items critical to the success of the project.

The work products that support the tracking of critical dependencies are directly related to the three key inputs described above. All changes in the status of critical dependencies affect the project plan and must be reflected in the plan as plan updates. Corrective action, and the documentation of lessons learned, may be triggered as a result of status reporting. Any corrective action, and lessons learned, may be recorded as the resolution of status report action items. Requested changes may be recorded any number of ways. These may be recorded as notes from a status review, or, more formally, as part of a change request process.

For integrated project management with other impacting projects, several planning and management tasks must be enhanced for interproject communications. Table C-28 details tasks specifically for IPM SP 2.2—Manage Dependencies.

RISK MANAGEMENT

Risk Taxonomy

The taxonomy in Figure C-37 was developed by the SEI [48] in support of risk management process development.

Table C-28. Tasks in support of integrated project management

Critical Dependencies Tracking—Detailed Project Planning
1. Detail project definition and work plan narrative and dependences on other projects
2. Define work-breakdown structure (WBS) and dependence on other project plan's WBS
3. Develop list of relevant stakeholders, including other projects stakeholders
4. Develop communication plan (internal, with other projects, and external)
5. Layout task relationships (proper sequence: dependencies, predecessor and successor tasks)
6. Identify critical dependencies
7. Estimate timelines
8. Determine the planned schedule and identify milestones
9. Determine the planned schedule and identify common milestones with other projects
10. Obtain stakeholder (including other projects) buyoff and signoff
11. Obtain Executive Approval to Continue

Critical Dependencies Tracking—Detailed Project Plan Execution
12. Mark milestones (significant event markers) and critical dependencies
13. Assign and level resources against shared resources pool
14. Set project baseline
15. Set up issues log
16. Set up change request log
17. Set up change of project scope log
18. Record actual resource usage and costs, including other projects
19. Track and manage critical path and critical dependencies
20. Manage and report status as determined by project plan
21. Communicate with stakeholders, including other projects, as determined by project plan
22. Report monthly project status, including other projects
23. Close project by formal acceptance of all stakeholders
24. Report on project actuals versus baseline estimates
25. Complete post implementation review report—lessons learned
26. Complete and file all project history documentation in document repository
27. Celebrate

Risk Taxonomy Questionnaire

The risk taxonomy questionnaire (Table C-29) was developed by the SEI [48] in support of risk management process development. This questionnaire can be extremely helpful when trying to identify and categorize project risk.

Risk Action Request

Table C-30. provides the suggested content for a risk action request that may be used to effectively support the assessment of project risk. The recommended content shown is based on the requirements found in IEEE Std 1540-2001, IEEE Standard for Software Lifecyle Processes—Risk Management, Annex B; Risk Action Request.

Risk Mitigation Plan

Table C-31 provides suggested content for a risk mitigation plan that may be used to document and control project risk. The recommended content shown is based on the require-

A. Product Engineering	B. Development Environment	C. Program Constraints

1. Requirements
  a. Stability
  b. Completeness
  c. Clarity
  d. Validity
  e. Feasibility
  f. Precedent
  g. Scale
2. Design
  a. Functionality
  b. Difficulty
  c. Interfaces
  d. Performance
  e. Testability
  f. Hardware Constraints
  g. Non-Developmental
    Software
3. Code and Unit Test
  a. Feasibility
  b. Testing
  c. Coding/
    Implementation
4. Integration and Test
  a. Environment
  b. Product
  c. System
5. Engineering Specialties
  a. Maintainability
  b. Reliability
  c. Safety
  d. Security
  e. Human Factors
  f. Specifications
	1. Development Process
	  a. Formality
	  b. Suitability
	  c. Process Control
	  d. Familiarity
	  e. Product Control
	2. Development System
	  a. Capacity
	  b. Suitability
	  c. Usability
	  d. Familiarity
	  e. Reliability
	  f. System Support
	  g. Deliverability
	3. Management Process
	  a. Planning
	  b. Project Organization
	  c. Management
	    Experience
	  d. Program Interfaces
	4. Management Methods
	  a. Monitoring
	  b. Personnel
	    Management
	  c. Quality Assurance
	  d. Configuration
	    Management
	5. Work Environment
	  a. Quality Attitude
	  b. Cooperation
	  c. Communication
	  d. Morale
		1. Resources
		  a. Schedule
		  b. Staff
		  d. Budget
		  d. Facilities
		2 Contract
		  a. Type of contract
		  b. Restrictions
		  c. Dependencies
		3. Program Interfaces
		  a. Customer
		  b. Associate Contractors
		  c. Subcontractor
		  d. Prime Contractor
		  e. Corporate
		    Management
		  f. Vendors
		  g. Politics

Figure C-37. Risk taxonomy

ments found in IEEE Std 1540-2001, IEEE Standard for Software Lifecyle Processes—Risk Management, Annex C; Risk Treatment Plan.

Risk Matrix Sample

A risk matrix is a helpful tool when used to support risk analysis activities [76]. Table C-32 provides an example of a risk matrix. Tables C-33 and C-34 provide description of the associated severity and probability levels.

DECISION ANALYSIS AND RESOLUTION

Cost/Benefit Ratio

A cost/benefit ratio is a calculation that depicts the total financial return for each dollar invested. Many projects have ready access to staff "hours" rather than accounting figures,

Table C-29. Risk taxonomy questionnaire

A. Product Engineering
   Technical aspects of the work to be accomplished
   1. Requirements
      a. Stability
         Are requirements changing even as the product is being produced?
      b. Completeness
         Are requirements missing or incompletely specified?
      c. Clarity
         Are the requirements unclear or in need of interpretation?
      d. Validity
         Will the requirements lead to the product the customer has in mind?
      e. Feasibility
         Are there requirements that are technically difficult to implement?
      f. Precedent
         Do requirements specify something never done before or beyond the experience of program personnel?
      g. Scale
         Is the system size or complexity a concern?
   2. Design
      a. Functionality
         Are there any potential problems in designing to meet functional requirements?
      b. Difficulty
         Will the design and/or implementation be difficult to achieve?
      c. Interfaces
         Are internal interfaces (hardware and software) well defined and controlled?
      d. Performances
         Are there stringent response time or throughput requirements?
      e. Testability
         Is the product difficult or impossible to test?
      f. Hardware Constraints
         Does the hardware limit the ability to meet any requirements?
      g. Nondevelopmental Software
         Are there problems with software used in the program but not developed by the program?
   3. Code and Unit Test
      a. Feasibility
         Is the implementation of the design difficult or impossible?
      b. Testing
         Is the specified level and time for unit testing adequate?
      c. Coding/Implementation
         Are the design specifications in sufficient detail to write code? Will the design be changing while coding is being done?
   4. Integration and Test
      a. Environment
         Is the integration and test environment adequate? Are there problems developing realistic scenarios and test data to demonstrate any requirements?
      b. Product
         Is the interface definition inadequate, facilities inadequate, or time insufficient? Are there requirements that will be difficult to test?
      c. System
         Has adequate time been allocated for system integration and test? Is system integration uncoordinated? Are interface definitions or test facilities inadequate?

Table C-29. *Continued*

A. Product Engineering (*cont.*)
   5. Engineering Specialties
     a. Maintainability
       Will the implementation be difficult to understand or maintain?
     b. Reliability
       Are reliability or availability requirements allocated to the software? Will they be difficult to meet?
     c. Safety
       Are the safety requirements infeasible and not demonstrable?
     d. Security
       Are there unprecedented security requirements?
     e. Human Factors
       Is there any difficulty in meeting the human factor requirements?
     f. Specifications
       Is the documentation adequate to design, implement, and test the system?
B. Development Environment
  Methods, procedures, and tools in the production of the software products
   1. Development Process
     a. Formality
       Will the implementation be difficult to understand or maintain?
     b. Suitability
       Is the process suited to the development mode, e.g., spiral, prototyping? Is the development process supported by a compatible set of procedures, methods, and tools?
     c. Process Control
       Is the software development process enforced, monitored, and controlled using metrics?
     d. Familiarity
       Are the project members experienced in use of the process? Is the process understood by all project members?
     e. Product Control
       Are there mechanisms for controlling changes in the product?
   2. Development System
     a. Capacity
       Are there enough workstations and processing capacity for all the staff?
     b. Suitability
       Does the development system support all phases, activities, and functions of the program?
     c. Usability
       Do project personnel find the development system easy to use?
     d. Familiarity
       Have project personnel used the development system before?
     e. Reliability
       Is the system considered reliable?
     f. System Support
       Is there timely expert or vendor support for the system?
     g. Deliverability
       Are the definition and acceptance requirements defined for delivering the system to the customer?
   3. Management Process
     a. Planning
       Is the program managed according to a plan?
     b. Project Organization
       Is the program organized effectively? Are the roles and reporting relationships well defined?

(*continued*)

Table C-29. *Continued*

B. Development Environment (*cont.*)
  3. Management Process (*cont.*)
    c. Management Experience
      Are the managers experienced in software development, software management, the application domain, and the development process?
    d. Program Interfaces
      Is there a good interface with the customer and is the customer involved in decisions regarding functionality and operation?
  4. Management Method
    a. Monitoring
      Are management metrics defined and is development progress tracked?
    b. Personnel Management
      Are project personnel trained and used appropriately?
    c. Quality Assurance
      Are there adequate procedures and resources to assure product quality?
    d. Configuration Management
      Are the change procedures or version control, including installation site(s) adequate?
  5. Work Environment
    a. Quality Attitude
      Does the project lack orientation toward quality work?
    b. Cooperation
      Does the project lack team spirit? Does conflict resolution require management intervention?
    c. Communication
      Does the project lack awareness of mission or goals? Is, communication of technical information among peers and managers adequate?
    d. Morale
      Is there a nonproductive, noncreative atmosphere? Does the project lack rewards or recognition for superior work?
C. Program Constraints
  Methods, procedures, and tools in the production of the software products
  1. Resources
    a. Schedule
      Is the project schedule inadequate or unstable?
    b. Staff
      Is the staff inexperienced, lack domain knowledge, lack skills, or not inadequately sized?
    c. Budget
      Is the funding insufficient or unstable?
    d. Facilities
      Are the facilities inadequate for building and delivering the product?
  2. Contract
    a. Type of contract
      Is the contract type a source of risk to the project?
    b. Restrictions
      Does the contract include any inappropriate restrictions?
    c. Dependencies
      Does the program have any critical dependencies on outside products or services?
  3. Program Interfaces
    a. Customer
      Are there any customer problems such as a lengthy document-approval cycle, poor communication, or inadequate domain expertise?

Table C-29. *Continued*

C. Program Constraints (*cont.*)
   3. Program Interfaces (*cont.*)
      b. Associate Contractors
         Are there any problems with associate contractors such as inadequately defined or unstable interfaces, poor communication, or lack of cooperation?
      c. Subcontractor
         Is the program dependent on subcontractors for any critical areas?
      d. Prime Contractor
         Is the program facing difficulties with its prime contractor?
      e. Corporate Management
         Is there a lack of support or micromanagement from upper management?
      f. Vendors
         Are vendors unresponsive to program needs?
      g. Politics
         Are politics causing a problem for the program?

Table C-30. Risk action request suggested content

Unique Identifier
Date of Issue and Status
Approval Authority
Scope
Request Originator
Risk Category
Risk Threshold
Project Objectives
Project Assumptions
Project Constraints
Risk Description
Risk Liklihood and Timing
Risk Consequences
Risk Mitigation Alternatives
   Descriptions
   Recommendation
   Justificaton
Disposition

Table C-31. Risk mitigation plan suggested content

Unique Identifier
Cross-Reference to Risk Action Request Unique Identifier
Date of Issue and Status
Approval Authority
Scope
Request Originator
Planned Risk Mitigation Activities and Tasks
Performers
Risk Mitigation Schedule (including resource allocation)
Mitigation Measures of Effectiveness
Mitigation Cost and Impact
Mitigation Plan Management Procedures

Table C-32. Sample risk matrix

Probability Severity	Frequent	Probable	Occasional	Remote	Improbable
Catastrophic	IN	IN	IN	H	M
Critical	IN	IN	H	M	L
Serious	H	H	M	L	T
Minor	M	M	L	T	T
Negligible	M	L	T	T	T

Legend: T = Tolerable, L = Low, M = Medium, H = High, IN = Intolerable.

Table C-33. Description of severity levels

Severity	Consequence
Catastrophic	Greater than 6 month slip in schedule; greater than 10% cost overrun; greater than 10% reduction in product functionality
Critical	Less than 6 month slip in schedule; less than 1-% cost overrun; less than 10% reduction in product functionality
Serious	Less than 3 month slip in schedule; less than 5% cost overrun; less than 5% reduction in product functionality
Minor	Less than 1 month slip in schedule; less than 2% cost overrun; less than 2% reduction in product functionality

Table C-34. Description of probability levels

Probability	Description
Frequent	Anticipate occurrence several times a year (>10 events)
Probable	Anticipate occurrence repeatedly during the year (2 to 10 events)
Occasional	Anticipate occurrence some time during the year (1 event)
Remote	Occurrence is unlikely though conceivable (< 1 event per year)

Table C-35. Cost/benefit ratio calculation

	Benefit (first year hours)	Total investment (hours)	Cost/benefit ratio	Return after 1 year for each dollar invested
Planned	6,400	1,000	6.4	> $6
Actual	6,000	800	7.5	> $7

which is good enough for decision analysis and resolution. The simplified example that follows is based on a fictitious case of developing a Web-based user-support program that is estimated to cost 1,000 hours to develop and deliver, and which should save $6,400 staff hours in the first year due to a reduction in field-service calls. Table C-35 describes the calculation.

REFERENCES

IEEE PUBLICATIONS

[1] IEEE/ANSI. *IEEE Guide to Software Configuration Management*. ANSI/IEEE Std 1042-1987, IEEE Press, New York. 1987. 92 pages.

[2] *IEEE Standard Glossary of Software Engineering Terminology*. IEEE Std 610.12-1990 (Sept 28), Reaffirmed Sept 2002, IEEE Press, New York. 2002.

[3] *IEEE Standard for Software Quality Assurance Plans*. IEEE Std 730-2002 (Sept), IEEE Press, New York. 2002.

[4] *IEEE Standard for Software Configuration Management Plans*. IEEE Std 828-1998 (Jun 25), IEEE Press, New York. 1998.

[5] IEEE Standard for Software Test Documentation, IEEE Std 829-1998 (Sep 16), IEEE Press, New York. 1998.

[6] *IEEE Recommended Practice for Software Requirements Specifications*. IEEE Std 830-1998 (Jun 25), IEEE Press, New York. 1998.

[7] *IEEE Standard Dictionary of Measures to Produce Reliable Software*. IEEE Std 982.1-1988 (Jun 9), IEEE Press, New York. 1988.

[8] An American National Standard—IEEE Standard for Software Unit Testing. ANSI/IEEE Std 1008-1987(R1993), Reaffirmed Dec 2002, IEEE Press, New York. 2002.

[9] *IEEE Standard for Software Verification and Validation*. IEEE Std 1012-1998 (Mar 9), IEEE Press, New York. 1998.

[10] *Supplement to IEEE Standard for Software Verification and Validation: Content Map to IEEE/EIA 12207.1-1996*. IEEE Std 1012a-1998 (Sept 16), IEEE Press, New York. 1998.

[11] *IEEE Recommended Practice for Software Design Descriptions*. IEEE Std 1016-1998 (Sept 23), IEEE Press, New York. 1998.

[12] *IEEE Standard for Software Reviews*. IEEE Std 1028-1997 (Mar 4), Reaffirmed Sept. 2002, IEEE Press, New York. 2002.

[13] *IEEE Standard Classification for Software Anomalies*. IEEE Std 1044-1993 (Dec 2), Reaffirmed Sept. 2002, IEEE Press, New York. 2002.

[14] *IEEE Standard for Software Productivity Metrics*. IEEE Std 1045-1992 (Sept 17), Reaffirmed Dec 2002, IEEE Press, New York. 2002.

[15] *IEEE Standard for Software Project Management Plans*. IEEE Std 1058-1998 (Dec 8), IEEE Press, New York. 1998.

[16] *IEEE Standard for a Software Quality Metrics Methodology*. IEEE Std 1061-1998 (Dec 8), IEEE Press, New York. 1998.

[17] *IEEE Recommended Practice for Software Acquisition*. IEEE Std 1062-1998 Edition (Dec 2), Reaffirmed Sept 2002, IEEE Press, New York. 2002.

[18] *IEEE Standard for Software User Documentation*. IEEE Std 1063-2001 (Dec 5), IEEE Press, New York. 2001.

[19] *IEEE Standard for Developing Software Life Cycle Processes*. IEEE Std 1074-1997 (Dec 9), IEEE Press, New York. 1997.

[20] *IEEE Standard Reference Model for Computing System Tool Interconnections*. IEEE Std 1175-1991 (Dec 5), IEEE Press, New York. 1991.

[21] *IEEE Guide for CASoftware Engineering Tool Interconnections—Classification and Description*. IEEE Std 1175.1-2002 (Nov. 11), IEEE Press, New York. 2002.

[22] *IEEE Standard for Software Maintenance*. IEEE Std 1219-1998 (Jun 25), IEEE Press, New York. 1998.

[23] *IEEE Standard for the Application and Management of the Systems Engineering Process*. IEEE Std 1220-1998 (Dec 8), IEEE Press, New York. 1998.

[24] *IEEE Standard for Software Safety Plans*. IEEE Std 1228-1994 (Mar 17), Reaffirmed Dec. 2002, IEEE Press, New York. 2002.

[25] *IEEE Guide for Developing System Requirements Specifications*. IEEE Std 1233, 1998 Edition (Apr 17), Reaffirmed Sept 2002, IEEE Press, New York. 2002.

[26] *IEEE Standard for Functional Modeling Language—Syntax and Semantics for IDEF0*. IEEE Std 1320.1-1998 (Jun 25), IEEE Press, New York. 1998.

[27] *IEEE Standard for Conceptual Modeling Language Syntax and Semantics for IDEF1X 97 (IDEF Object)*. IEEE Std 1320.2-1998 (Jun 25), IEEE Press, New York. 1998.

[28] *IEEE Guide for Information Technology—System Definition—Concept of Operations (ConOps) Document*. IEEE Std 1362-1998 (Mar 19), IEEE Press, New York. 1998.

[29] *IEEE Standard for Information Technology—Software Reuse—Data Model for Reuse Library Interoperability: Basic Interoperability Data Model (BIDM)*. IEEE Std 1420.1-1995 (Dec 12), Reaffirmed Jun 2002, IEEE Press, New York. 2002.

[29a] Supplement to IEEE Standard for Information Technology—Software Reuse—Data Model for Reuse Library Interoperability: Asset Certification Framework. IEEE Std 1420.1a-1996 (Dec 10), Reaffirmed Jun 2002, IEEE Press, New York. 2002.

[30] *IEEE Trial-Use Supplement to IEEE Standard for Information Technology—Software Reuse—Data Model for Reuse Library Interoperability: Intellectual property Rights Framework*. IEEE Std 1420.1b-1999 (Jun 26), Reaffirmed Jun 2002, IEEE Press, New York. 2002.

[31] *IEEE Standard—Adoption of International Standard ISO/IEC 14102: 1995—Information Technology—Guideline for the Evaluation and Selection of CASoftware Engineering Tools*. IEEE Std 1462-1998 (Mar 19), IEEE Press, New York. 1998.

[32] *IEEE Standard—Adoption of International Standard ISO/IEC 12119: 1994(E)—Information Technology—Software Packages—Quality Requirements and Testing*. IEEE Std 1465-1998 (Jun 25), IEEE Press, New York. 1998.

[33] *IEEE Recommended Practice for Architectural Description of Software-Intensive Systems*. IEEE Std 1471-2000 (Sept 21), IEEE Press, New York. 2000.

[34] *IEEE Guide—Adoption of PMI Standard—A Guide to the Project Management Body of Knowledge.* IEEE Std 1490-2003 (Dec 10), Replaces 1490-1998 (Jun 25), IEEE Press, New York. 2003.

[35] *EIA/IEEE Interim Standard for Information Technology—Software Life Cycle Processes— Software Development: Acquirer–Supplier Agreement.* IEEE Std 1498-1995 (Sept 21), IEEE Press, New York. 1995.

[36] *IEEE Standard for Information Technology—Software Life Cycle Processes—Reuse Processes.* IEEE Std 1517-1999 (Jun 26), IEEE Press, New York. 1999.

[37] *IEEE Standard for Software Life Cycle Processes—Risk Management.* IEEE Std 1540-2001 (Mar 17), IEEE Press, New York. 2001.

[38] *IEEE Recommended Practice for Internet Practices—Web Page Engineering—Intranet/Extranet Applications.* IEEE Std 2001-2002 (Jan 21), IEEE Press, New York. 2003.

[39] *Industry Implementation of International Standard ISO/IEC 12207:1995—Standard for Information Technology—Software Life Cycle Processes.* IEEE/EIA 12207.0/.1/.2-1996 (Mar), IEEE Press, New York. 1996.

[40] IEEE. *IEEE Standards Collection, Software Engineering,* 1994 Edition. IEEE Press, New York. 1994.

[41] IEEE. *IEEE Software Engineering Standards Collection.* IEEE Press, New York. 2003.

[42] IEEE. Software and Systems Engineering Standards Committee Charter Statement, http://standards.computer.org/S2ESC/S2ESC_pols/S2ESC_Charter.htm. 2003.

[43] S2ESC Guide for Working Groups, http://standards.computer.org/S2ESC/S2ESC_wgresources/ S2ESC-WG-Guide-2003-07-14.doc. 2003.

[44] IEEE. *Guide to the Software Engineering Body of Knowledge (SWEBOK),* Trial Version. IEEE Press, New York. 2001.

[45] McConnell, S., "The Art, Science, and Engineering of Software Development," *IEEE Software Best Practices, 15,* 1, 1998.

SEI PUBLICATIONS

[46] Bothwell, C. and Masters, S., "CMM® Appraisal Framework Version 1.0," Software Engineering Institute, Carnegie Mellon University, Technical Report, CMU/SEI-95-TR-001. 1995.

[47] Bounds, N. M., and Dart S. A., *Configuration Management (CM) Plans: The Beginning to your CM Solution,* Software Engineering Institute, Carnegie Mellon University. 1993.

[48] Carr, Marvin et al., *Taxonomy-Based Risk Identification,* Software Engineering Institute, Carnegie Mellon University, Technical Report, CMU/SEI-93-TR-006. 1993.

[49] *The Capability Maturity Model: Guidelines for Improving the Software Process,* v.1.1, Software Engineering Institute, Carnegie Mellon University. 1997.

[50] Draper, G. and Hefner, R., *Applying CMMI®* Generic Practices with Good Judgment, SEPG Conference Tutorial, http://www.sei.cmu.edu/cmmi/presentations/sepg04.presentations/apply-gps.pdf. 2004.

[51] Dunaway, D. K., *CMM®* Based Appraisal for Internal Process Improvement (CBA IPI). Team Member's Handbook, v1.1, Software Engineering Institute, Carnegie Mellon University, Handbook, CMU/SEI-96-HB-005. 1996.

[52] Dunaway, D.K. and Masters S., "CMM-Based Appraisal for Internal Process Improvement (CBA IPI) Method Description," Software Engineering Institute, Carnegie Mellon University, Technical Report, CMU/SEI-96-TR-007. 1996.

[53] Ford, G. and Gibbs, N., "A Mature Profession of Software Engineering," Software Engineering Institute, Carnegie Mellon University, Technical Report, CMU/SEI-96-TR-004. 1996.

[54] Gremba, J. and Myers, C., "The IDEAL Model: A Practical Guide for Improvement," *Bridge,* Issue 3, Software Engineering Institute, Carnegie Mellon University, 1997.

[55] Kasunic, Mark, "An Integrated View of Process and Measurement," Software Engineering Institute, Carnegie Mellon University, Presentation, http://www.sei.cmu.edu/sema/pdf/integrated-view-process.pdf. 2004.

[56] McFeeley, B., *IDEAL: A User's Guide to Software Process Improvement,* Software Engineering Institute, Carnegie Mellon University, Handbook, CMU/SEI-96-HB-001. 1996.

[57] Paulk, M. C., Weber, C. V., Garcia, S. M., Chrissis, M. B., and Bush, M., *Key Practices of the Capability Maturity Model, Version 1.1,* Software Engineering Institute, Carnegie Mellon University. 1993.

[58] Phifer, E., *DAR Basics: Applying Decision Analysis and Resolution in the Real World,* Software Engineering Institute, Carnegie Mellon University, SEPG Presentation, http://www.sei.cmu.edu/cmmi/presentations/sepg04.presentations/dar.pdf. 2004.

[59] SEI, *A Framework for Software Product Line Practice,* V 4.2, Software Engineering Institute, Carnegie Mellon University, Web Report, http://www.sei.cmu.edu/plp/framework.html#outline. 2004.

[60] SEI, *Appraisal Requirements for CMMI®* Version 1.1 (ARC, V1.1), Software Engineering Institute, Carnegie Mellon University, Technical Report, CMU/SEI-2001-TR-034. 2001.

[61] SEI, *Capability Maturity Model Integration (CMMI) for Software Engineering,* v.1.1 Staged Representation, Software Engineering Institute, Carnegie Mellon University, Technical Report, CMU/SEI-2002-TR-029. 2002.

[62] SEI, *CMM®* Based Appraisal: Team Training Participant's Guide, CMM® Based Appraisal for Internal Process Improvement (CBA IPI), Software Engineering Institute, Carnegie Mellon University. 1998.

[63] SEI, *Component Screening Criteria and Rationale Guideline,* Software Engineering Institute, Carnegie Mellon University, EPIC Artifact. 2004.

[64] SEI, *Integrated Product Development (IPD)-CMM®,* Software Engineering Institute, Carnegie Mellon University, Model Draft. 1997.

[65] SEI, *Organizational Process Improvement Checklist Using IDEAL^{SM}*, Software Engineering Insititute, Carnegie Mellon University, Presentation, www.sei.cmu.edu/ideal/ideal.present, 2004.

[66] SEI, *Standard CMMI®* Appraisal Method for Process Improvement (SCAMPI), V1.1: Method Definition Document, Software Engineering Institute, Carnegie Mellon University. 2001.

[67] Zubrow, D., Hayes, W., Siegel, J., and Goldenson, D., *Maturity Questionnaire,* Special Report, CMU/SEI-94-SR-7, Software Engineering Institute, Carnegie Mellon University. 1994.

OTHER REFERENCES

[68] Arthur, L., *Software Evolution: The Software Maintenance Challenge,* Wiley. 1988.

[69] Babich, W., *Software Configuration Management,* Addison-Wesley. 1986.

[70] Victor R. Basili et al., "A Reference Architecture for the Component Factory," *ACM Transactions on Software Engineerig and Methodology, 1.,* 1, 53–80, Jan. 1992.

[71] Bersoff, E., Henderson, V., and Siegel, S., *Software Configuration Management: A Tutorial,* pp. 24–32, IEEE Computer Society Press. 1980.

[72] Bredemeyer Consulting, *The Architecture Discipline—Software Architecting Success Factors and Pitfalls,* http://www.bredemeyer.com/CSFs_pitfalls.htm. 2004.

[73] Croll, P., "Eight Steps to Success in CMMI-Compliant Process Engineering, Strategies and Supporting Technology," in *Third Annual CMMI® Technology Conference and Users Group*. 2003.

[74] Croll, P., "How to Use Standards as Best Practice Information Aids for CMMI-Compliant Process Engineering," in *14th Annual DoD Software Technology Conference*. 2002.

[75] Davis, A., *Software Requirements: Analysis and Specification,* Prentice-Hall. 1990.

[76] U.S. Department of Defense, *Software Transition Plan,* Data Item Description DI-IPSC-81429.

[77] U.S. Department of Energy Quality Managers Software Quality Assurance Subcommittee, "Software Risk Management, A Practical Guide," SQAS21.01.00—1999. 2000.

[78] U.S. Department of Justice Systems Development Life Cycle Guidance, Interface Control Document Template, Appendix C-17; http://www.usdoj.gov/jmd/irm/lifecycle/table.htm, 2005.

[79] Dunn, R. H. and Ullman, R. S., *TQM for Computer Software,* 2nd ed., McGraw-Hill. 1994.

[80] Dymond, K. M., *A Guide to the CMM, Understanding the Capability Maturity Model for Software,* Process Transition International. 1998.

[81] EIA/IS 731, *Systems Engineering Capability Model,* Electronic Industries Alliance. 1994.

[82] Freedman, D. P., and Weinberg, G. M., *Handbook of Walkthroughs, Inspections, and Technical Reviews, Evaluating Programs, Projects, and Products,* 3rd ed., Dorset House. 1990.

[83] Glazer, H., "Two Key Challenges for Implementing CMM," http://www.entinex. com/ImplementingCMMI_page1.cfm. 2004.

[84] Hass, A. M. J., *Configuration Management Principles and Practice,* Addison-Wesley. 2003.

[85] Hefner, R., "New CMMI® Requirements for Risk Management," *CrossTalk, The Journal of Defense Software Engineering,* February 2000.

[86] ISO/IEC JTC1/SC7, ISO/IEC 14143.1:1999, *Information technology—Software Measurement— Functional Size Measurement—Definition of Concepts,* Canada. 1998.

[87] ISO/IEC JTC1/SC7, ISO/IEC 15288:2002, *Systems Engineering—System Life Cycle Processes,* Canada. 2002.

[88] ISO/IEC JTC1/SC7, ISO/IEC 15939:2001, *Information Technology—Software Engineering— Software Measurement Process,* Canada. 2001.

[89] Jalote P., *An Integrated Approach to Software Engineering,* 2nd ed., Springer-Verlag. 1997.

[90] Land, S. K., "First User's of Software Engineering Standards Survey," IEEE Software and Systems Engineering Standards Committee (S2ESC). 1997.

[91] Land S. K., "Second User's of Software Engineering Standards Survey," IEEE Software and Systems Engineering Standards Committee (S2ESC). 1999.

[92] Land, S. K., "IEEE Standards User's Survey Results," In *ISESS '97 Conference Proceedings,* IEEE Press. 1997.

[93] Land, S. K., "Second IEEE Standards User's Survey Results," In *ISESS '99 Conference Proceedings,* IEEE Press. 1999.

[94] Land, S. K., *Jumpstart CMM®/CMMI® Software Process Improvement: Using IEEE Software Engineering Standards,* Wiley. 2004.

[95] McConnell, S., *Professional Software Development,* Addison-Wesley. 2004.

[96] Moore, J., "Increasing the Functionality of Metrics through Standardization," In *Conference on Developing Strategic I/T Metrics.* 1998.

[97] Moore, J., *Road Map to Software Engineering—A Standards Based Guide,* Wiley. 2005.

[98] Pressman, R., *Software Engineering,* McGraw-Hill. 1987.

[99] Royce, Walker, "CMM® vs. CMMI, From Conventional to Modern Software Management," *The Rational Edge.* 2002.

[100] Schach, S. R., *Classical and Object-Oriented Software Engineering,* 3rd ed., Irwin. 1993.

[101] USAF Software Technology Support Center (STSC), *CMM-SE/SW V1.1 to SW-CMM*® V1.1 Mapping. 2002.

[102] Veenendall, E.V., Ammerlaan, R., Hendriks, R., van Gensewinkel, V., Swinkels, R., and van der Zwan, M., "Dutch Encouragement; Test standards we use in our projects," *Professional Tester,* 16, October 2003.

[103] Westfall, L. L., "Seven Steps to Designing a Software Metric," BenchmarkQA, Whitepaper. 2002.

[104] Whitgift, D., *Methods and Tools for Software Configuration Management,* Wiley. 1991.

[105] Wiegers, K. E., *Creating A Software Engineering Culture,* Dorset House. 1996.

[106] Williams and Wegerson, *Evolving the SEPG to a CMMI World,* http://www.sei.cmu.edu/cmmi/presentations/sepg03.presentations/williams-wegerson.pdf. 2002.

INDEX

ABOUT THE AUTHORS

Susan K. Land

Susan Land is a Software Engineering Section Manager for Northrop Grumman/TASC in Huntsville, Alabama. She has participated in many software process improvement efforts, CMM® CAF-based appraisals, and CBA IPI self-assessments. Ms. Land is currently a member of the IEEE Computer Society Board of Governors, Chair of the IEEE Standards Advisory Board SAB, the IEEE Software and Systems Engineering Standards Committee (S2ESC), the Editorial Board for the IEEE Computer Society (CS) Software Engineering Online publication, the IEEE CS Professional Practices Committee, and the development team of the Strawman and Trial versions of the Software Engineering Body of Knowledge (SWEBOK®). She is an IEEE Certified Software Development Professional (CSDP). Ms. Land is also the author of *Jumpstart CMM®/CMMI® Software Process Improvement: Using IEEE Software Engineering Standards.*

John W. Walz

John Walz retired as Senior Manager, Supply Chain Management, Lucent Technologies. His 30-year career at Lucent and AT&T was highlighted by customer-focused and dynamic, results-oriented management and over 20 years of management/coaching experience. John held leadership positions in hardware and software development, engineering, quality planning, quality auditing, quality standards implementation, and strategic planning. He was also responsible for Lucent's Supply Chain Network and a valuable team member for Lucent's quality strategic planning. This work resulted in many locations achieving ISO 9001 and then TL 9000 registrations. These organizations achieved and shared quality and process improvement results at annual Lucent sharing rallies and external best practices conferences. John actively participated in both software engineering and quality system standards development and implementation. Mr. Walz also wrote the TL 9000 Chapter in ISO 9000 Handbook, 4th edition. John is an expert speaker in the Distinguished Visitor Program of the IEEE Computer Society. John earned his BSEE and MSEE degrees from Ohio State University.